Copyright © <2023> <marksparnon>

All rights reserved.

No part of this book may be reproduced or used in any manner without the prior written permission of the copyright owner, except for the use of brief quotations in a book review.

ISBN: 9780648608745

To Georgia and Scarlett,

I will help you with the Dot to Dot puzzles but that's where I draw the line.

Children are remarkable for their ability to quickly and easily learn massive amounts of new information. It is proposed from recent research that the key behind this ability to rapidly process is the neurotransmitter GABA. Furthermore, functional MRI have proven children activate different and more regions of their brains than adults when they perform word tasks[*].

[*]Washington University School of Medicine in St. Louis.

Why is it then if children's brains absorb information at a much faster rate than adults that they are so hard to teach at home? We have a theory…

"The average 4-year-old laughs 300 times a day. The average 40-year-old? Only four."

This common quote is hard to attribute to a single individual and trace its origin. However It is undeniable to a parent that children laugh longer and louder and are more distracted than adults.

The desire to have fun competes with their increased capability to learn. Imagine a world where you both win. They do their homework (tick parenting win) and they have fun (tick child win). This lead to our research and development of Learning through fun.

This series focuses on mathematics and is based upon the New Zealand Curriculum, Australian Schools Curriculum (Version 9) and Naplan testing 2022. Each book contains a series of fun Dot to Dot pictures, joke puzzles and colour by number activities.

– Index –

Puzzle 1	Dot to Dot
Puzzle 2	Dot to Dot
Puzzle 3	Dot to Dot
Puzzle 4	A Joke for you
Puzzle 5	Dot to Dot
Puzzle 6	Dot to Dot
Puzzle 7	Colour by Number
Puzzle 8	Dot to Dot
Puzzle 9	Dot to Dot
Puzzle 10	Dot to Dot
Puzzle 11	A Joke for you
Puzzle 12	Colour by Number
Puzzle 13	Colour by Number
Puzzle 14	Dot to Dot
Puzzle 15	Dot to Dot
Puzzle 16	A Joke for you
Puzzle 17	Dot to Dot
Puzzle 18	Colour by Number
Puzzle 19	Dot to Dot
Puzzle 20	Dot to Dot
Puzzle 21	Dot to Dot
Puzzle 22	Colour by Number
Puzzle 23	A Joke for you
Puzzle 24	Dot to Dot
Puzzle 25	Dot to Dot
Puzzle 26	Dot to Dot
Puzzle 27	Dot to Dot
Puzzle 28	Dot to Dot
Puzzle 29	Colour by Number
Answers	

PUZZLE 1

Solve the equations and connect the dots in order

MATHS PROBLEM	ANSWER
Start: 3 - 2	
10 + 7	
32 - 4	
10 - 7	
7, 9, ?, 13	
10 x 10	
half of 4	
? + 2 = 20	
6 + ? = 20	
10 + 10 + 6	
3 x 10	
4, ?, 10, 13	
2 x 5	
30 - 7	

PUZZLE 2

Solve the equations and connect the dots in order

MATHS PROBLEM	ANSWER
3 + ? = 10	
3,6,?,12	
5 + 6	
2 x 10	
20 +20 + 2	
4, 8, 12, ?	
70 - 3	
10 + 20 + 6	
Half of 100	

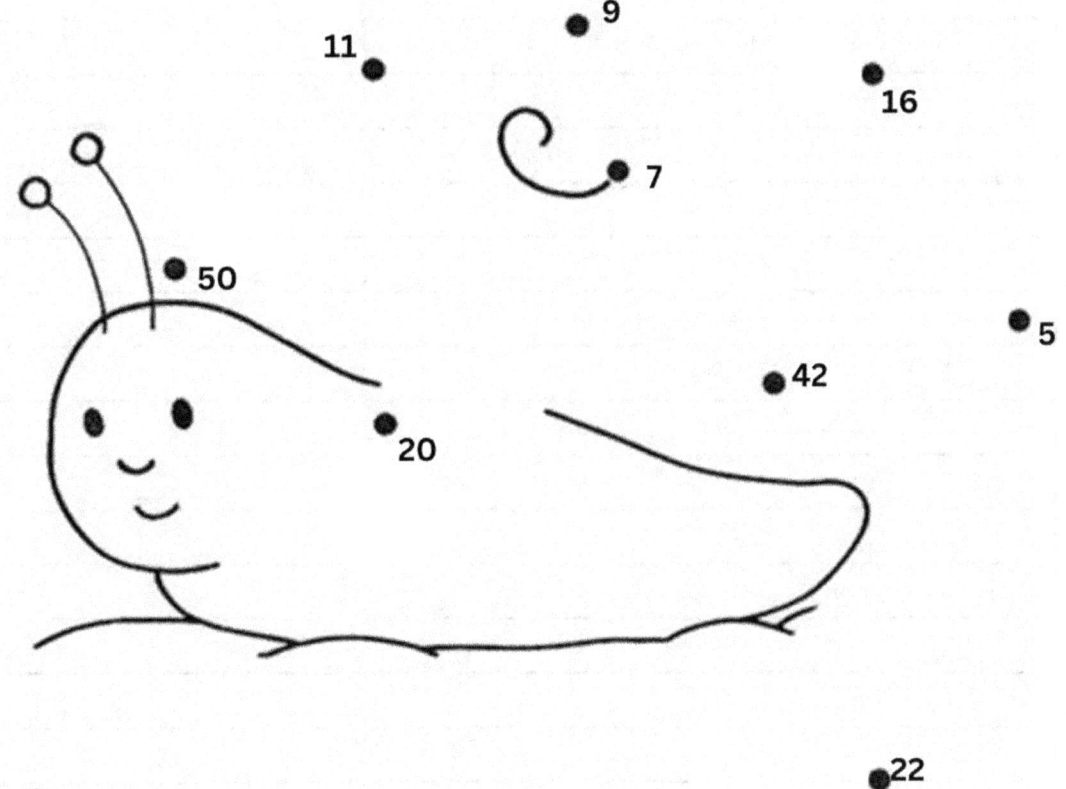

PUZZLE 3

Solve the equations and connect the dots in order

MATHS PROBLEM	ANSWER
START: 17 - 16	
6 +7	
30 + 30 + 30	
100, ? , 300, 400	
30 + 40 + 2	
2 x 4	
10 + ? = 13	
2 x 40	
37, 39, ?, 43	
50 - 6	
30 + 30	
50 + 5	
Half of 8	
10-8	
100 - 12	
11, 22, ?, 44	
20 +20 + 10 +7	
30 - 9	
Half of 10	
30 + 30 + 30 + 1	
How many minutes in a quarter of an hour?	
10 x 10	
8 + 8	
6+5	
30 - 10 - 19	

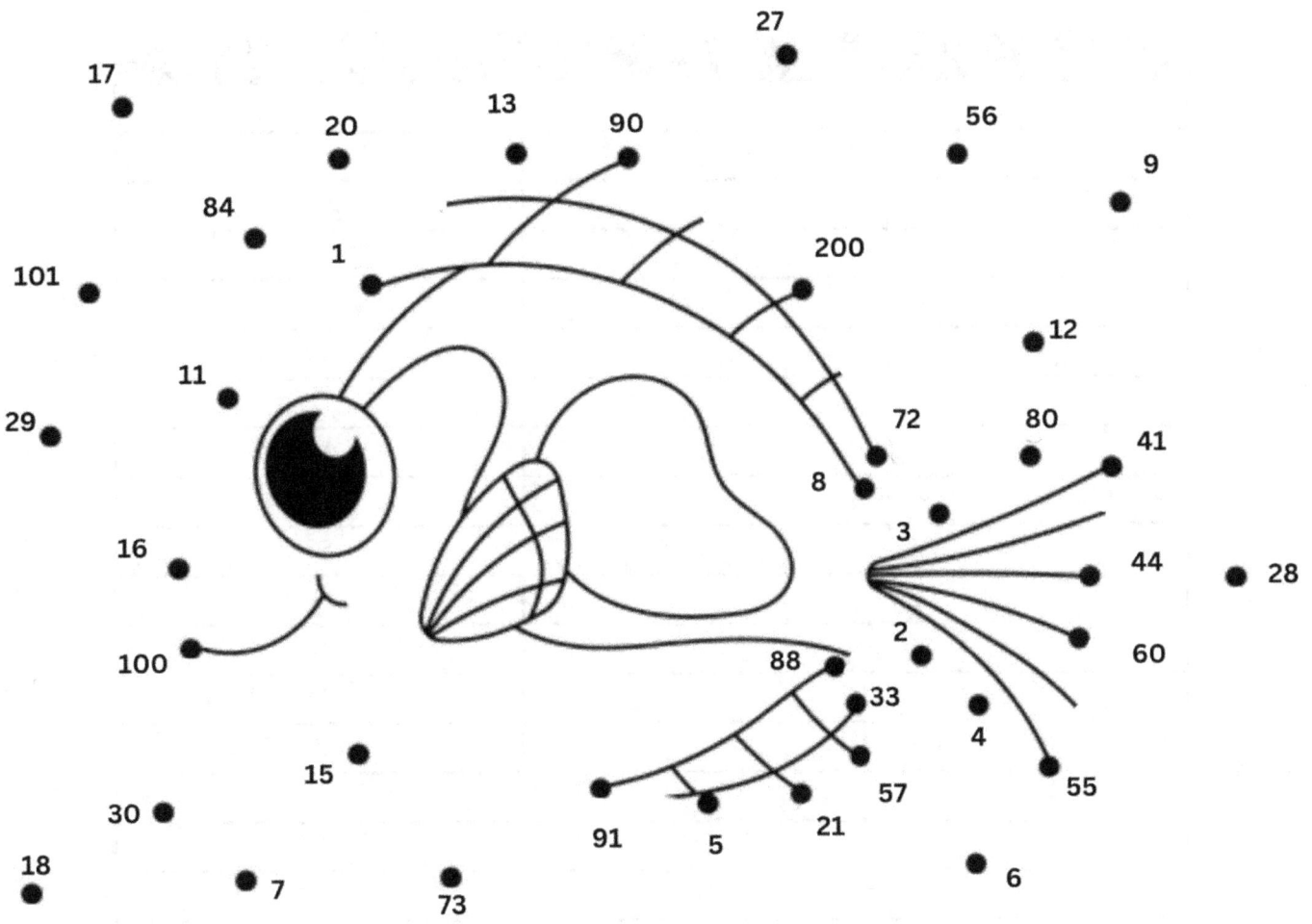

PUZZLE 4

Solve the equations to find the letters.

LETTER	ANSWER
A	3
B	21
C	200
D	100
E	2
F	50
G	5
H	12
I	29
J	56
K	4
L	6
M	25
N	20
O	14
P	150
Q	400
R	1000
S	10
T	7
U	46
V	36
W	110
X	240
Y	9
Z	68

How do you organize a space party?

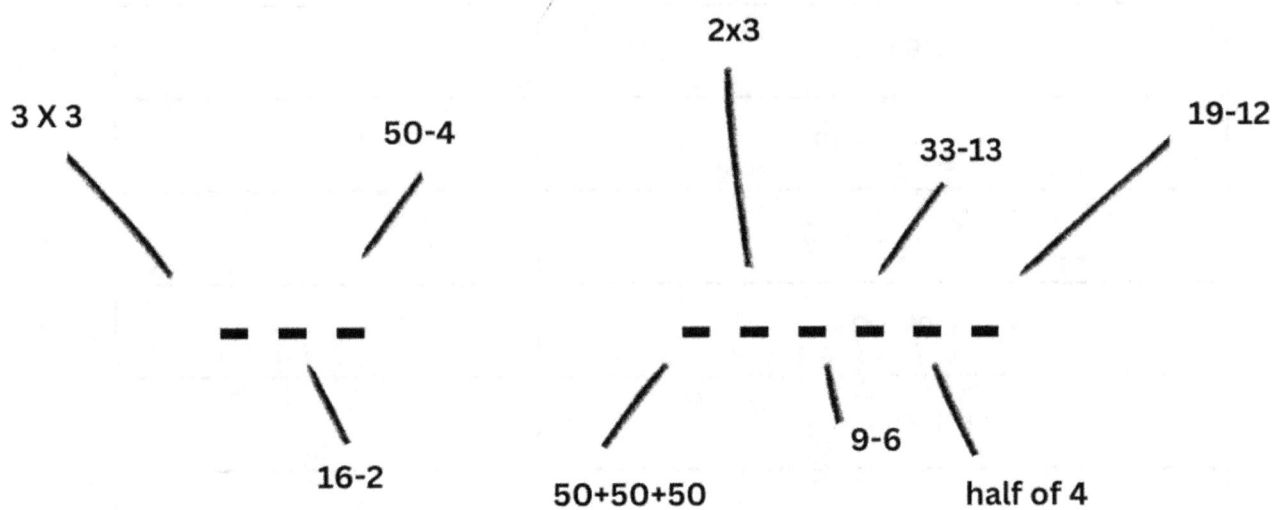

PUZZLE 5

Solve the equations and connect the dots in order

MATHS PROBLEM	ANSWER
20 + ? = 70	
90, 100, ?, 120	
11 x 2	
How many minutes in an hour?	
100 - 99	
50 + 50 +100	
10 + 10 + 20	
99 + 24	
101, 100, ?, 98	
100 - 10	
10 - 3	
7 + 8	

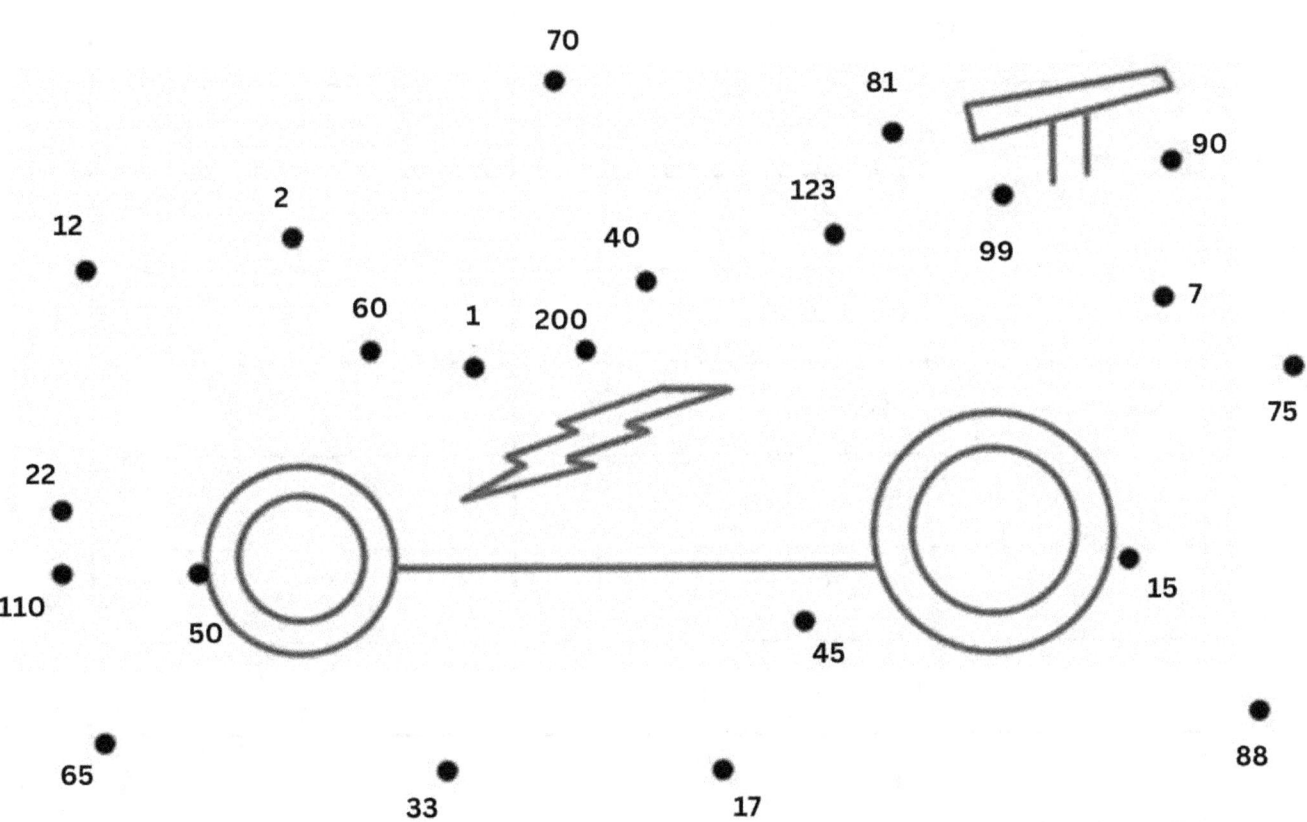

PUZZLE 6

Solve the equations and connect the dots in order

MATHS PROBLEM	ANSWER
10 +10 +10 + 7	
20 x 3	
10 +10 +7	
Half of 10	
53, 56, ?, 62	
15 - 5 - 9	
20 x 2	
6, ?, 26, 36	
40 -14	
1 + 9 +40	

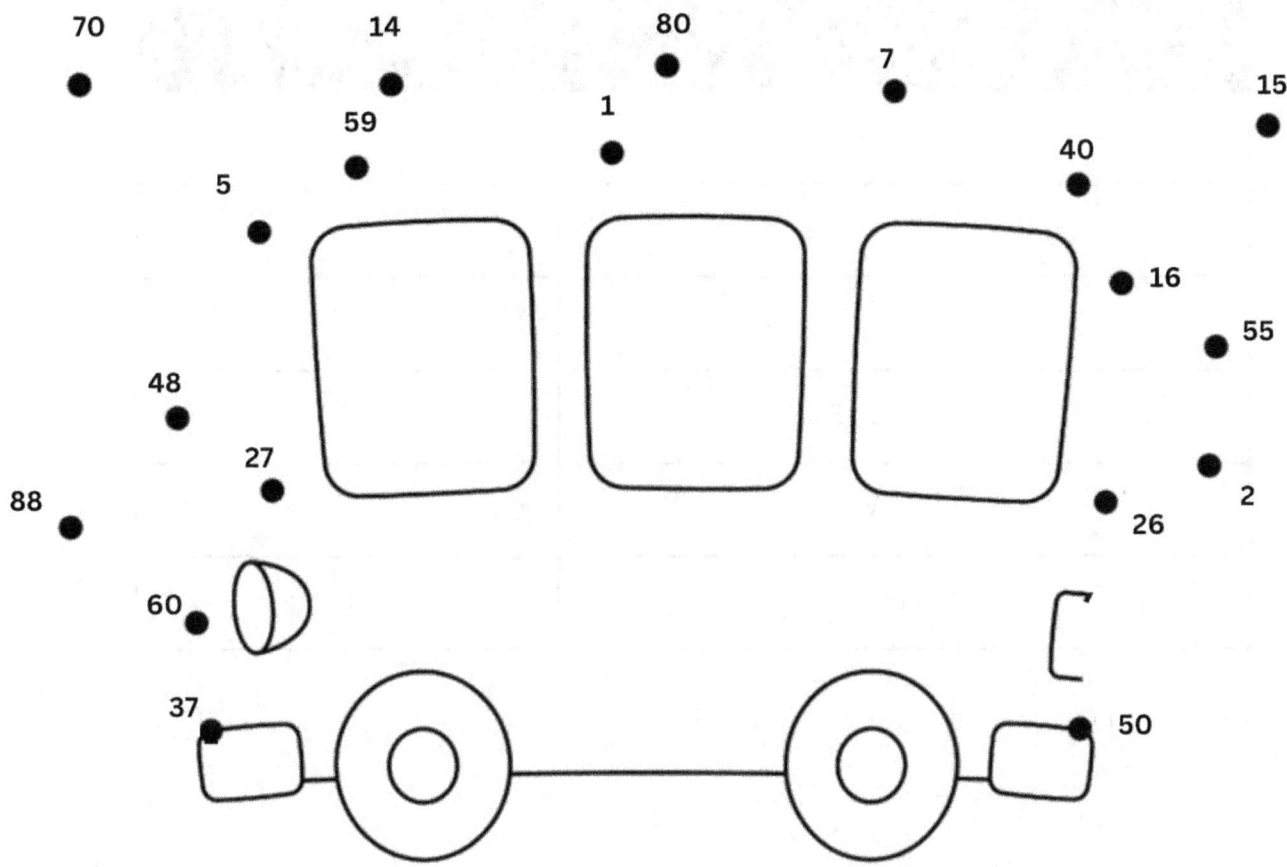

PUZZLE 7

Colour by Number

MATHS PROBLEM	ANSWER
BLACK	11
RED	5
GREY	3
LIGHT BLUE	4
YELLOW	2
ORANGE	6

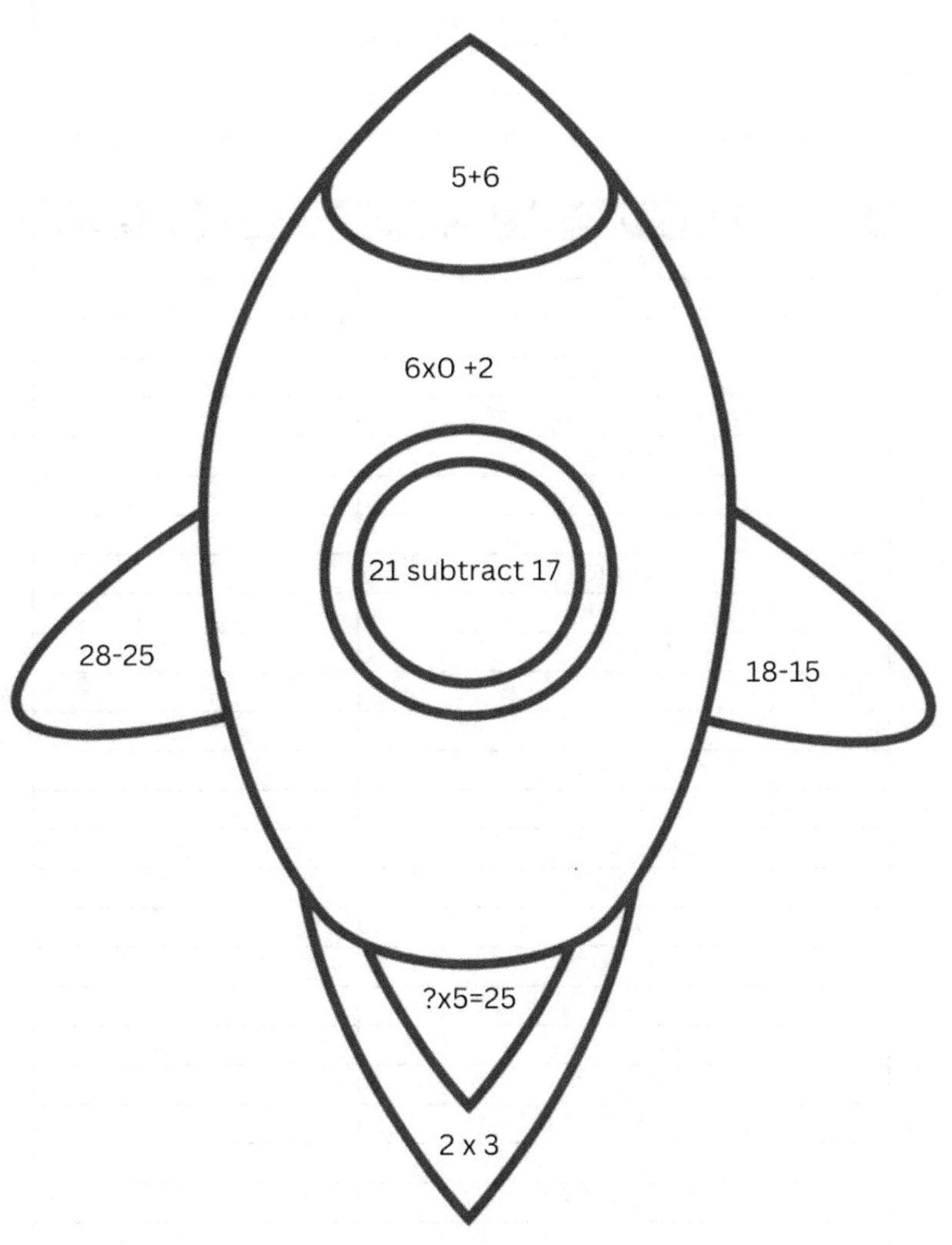

PUZZLE 8

Solve the equations and connect the dots in order

MATHS PROBLEM	ANSWER
13 - 12	
2 x 4	
8, ?, 4, 2	
100 + 200 + 100	
How many days in a fortnight?	
6 x ? = 12	
50 - 9	
80 - 3	
17 + 6	
7, 9, ?, 13	
11 x 2	
3 x 3	
80 -14	
15, ?, 9, 6	
? - 16 = 10	
20 + 30 + 4	
50 + 20 + 30	
3, 9, ?, 21	
5 x ? = 15	
90 -9	
50 + 10 + 1	
2000 -1000	
23 +10	
1, ?, 9, 13	
30 + 30 + 30 + 1	
50 + 50 + 6	
Half of 20	
3 x 5	

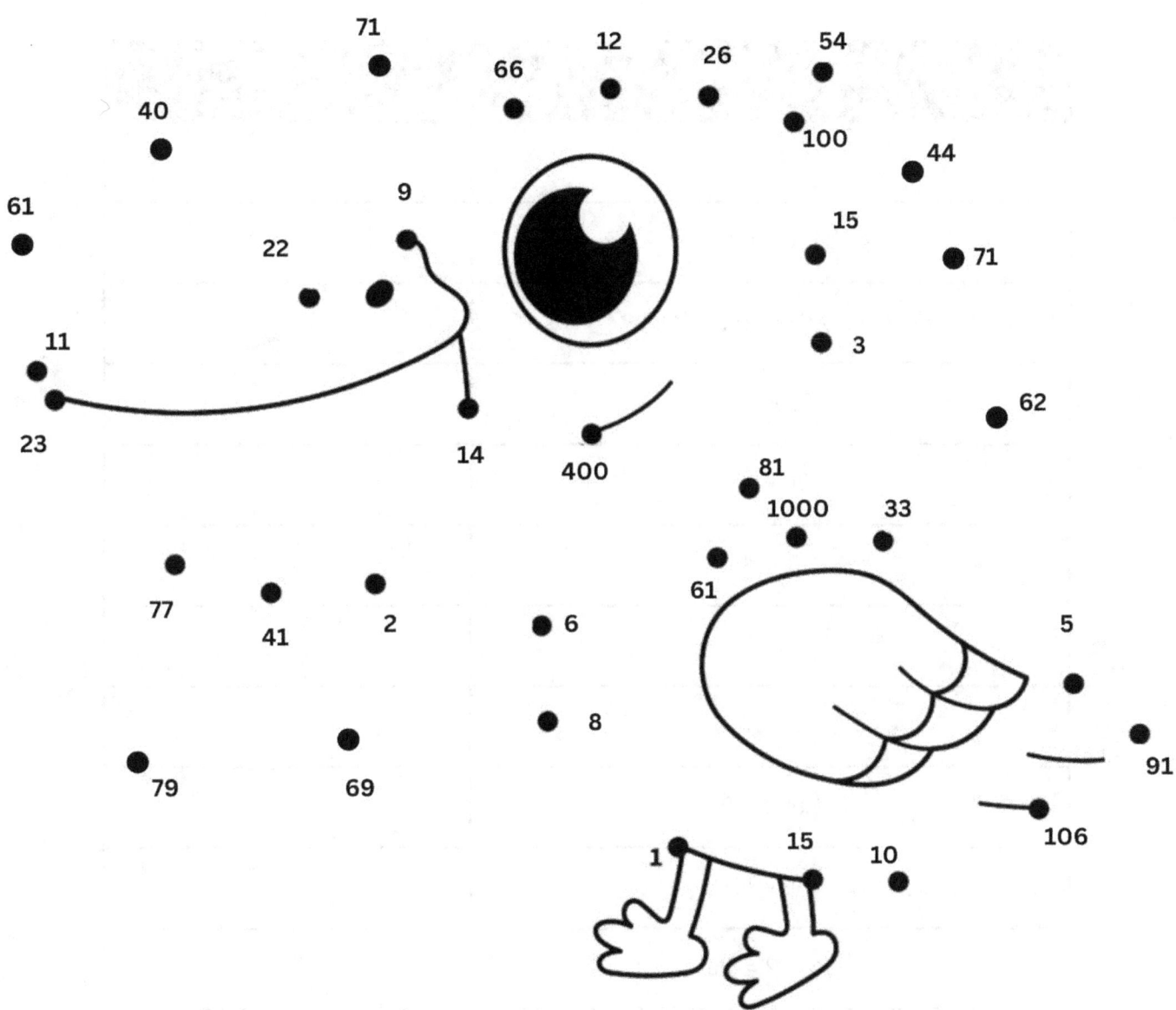

PUZZLE 9

Solve the equations and connect the dots in order

MATHS PROBLEM	ANSWER
1/4 of 4	
126 - 100	
How many days in a fortnight?	
51 + 20	
20 - 10	
3 + 4	
19 - 8	
6 x 10	
Half of 10	
40 - 20	
16 + ? = 20	
10 + 10 + 10 + 6	
10 x 10	

PUZZLE 10

Solve the equations and connect the dots in order

MATHS PROBLEM	ANSWER
START: 8 - 5 - 2	
101 -2	
3 x 4	
3 x 3	
10 - 3	
31 + 10 + 10	
20 -10	
21 - 2	
27, 37, ?, 57	
68 + 20	
23 = 22 + ?	

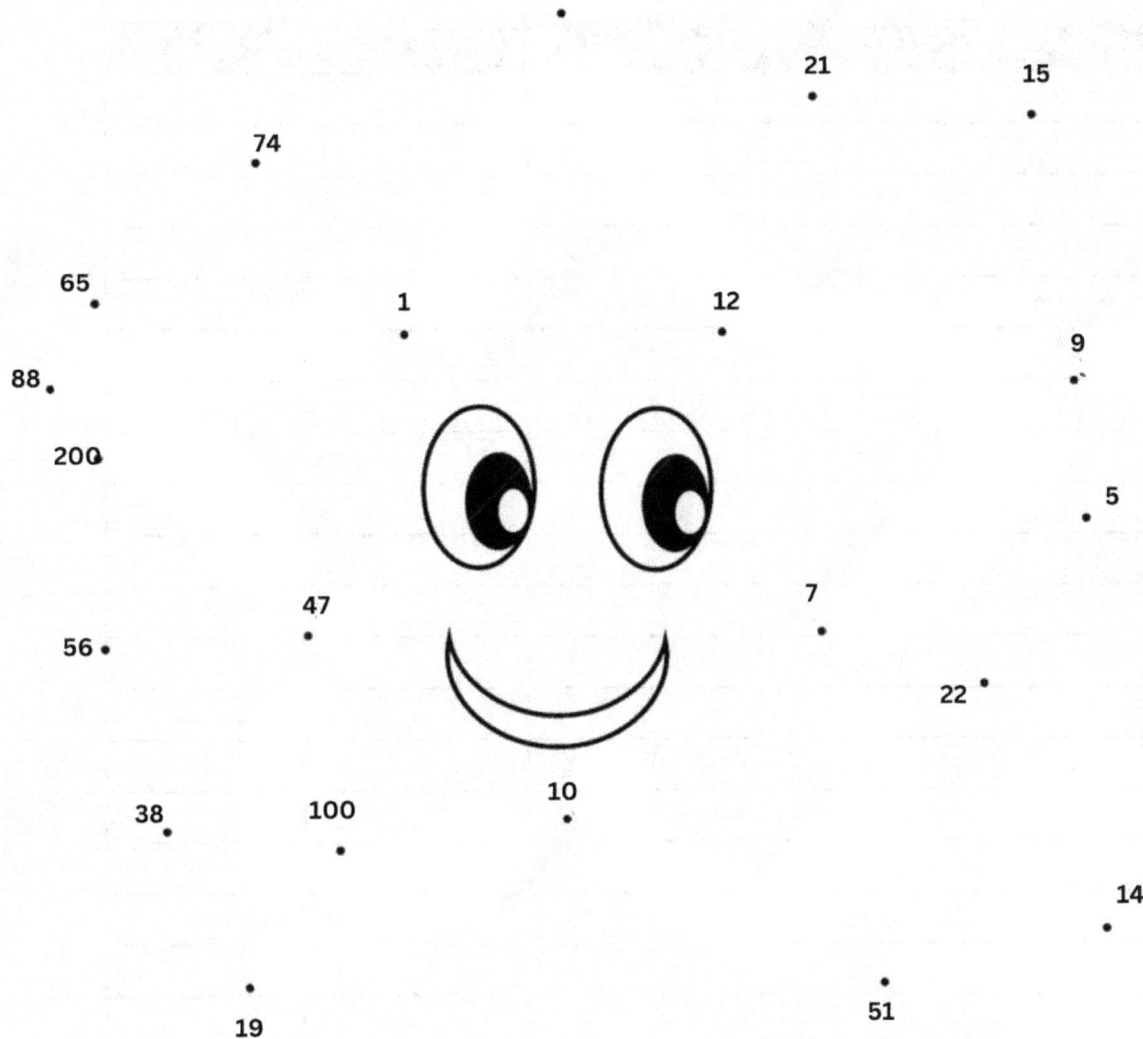

PUZZLE 11

Solve the equations to find the letters.

LETTER	ANSWER
A	3
B	21
C	200
D	100
E	2
F	50
G	5
H	12
I	29
J	56
K	4
L	6
M	25
N	20
O	14
P	150
Q	400
R	1000
S	10
T	7
U	46
V	36
W	110
X	240
Y	9
Z	68

What do you get when you cross a sheep with a kangaroo?

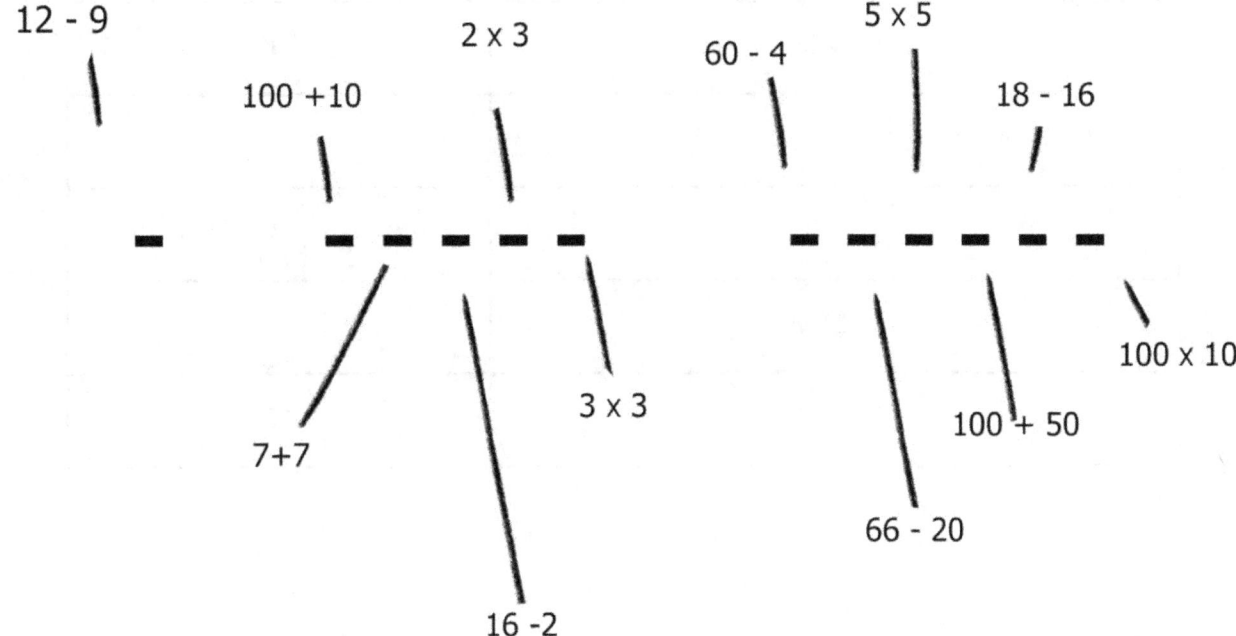

12 - 9   100 +10   2 x 3   60 - 4   5 x 5   18 - 16

___   _ _ _ _   _ _ _ _ _

    7+7   3 x 3   66 - 20   100 + 50   100 x 10
       16 -2

PUZZLE 12

Colour By Number

MATHS PROBLEM	ANSWER
BLACK	10
RED	6
DARK BLUE	1
ORANGE	7
SILVER	9
GREY	5
WHITE	3
YELLOW	2
PURPLE	12

PUZZLE 13

Colour By Number

MATHS PROBLEM	ANSWER
light blue	1, 8 or 11
yellow	3, 13 or 100
black	2, 20 or 34
orange	5, 9 or 87
pink	4, 51 or 68

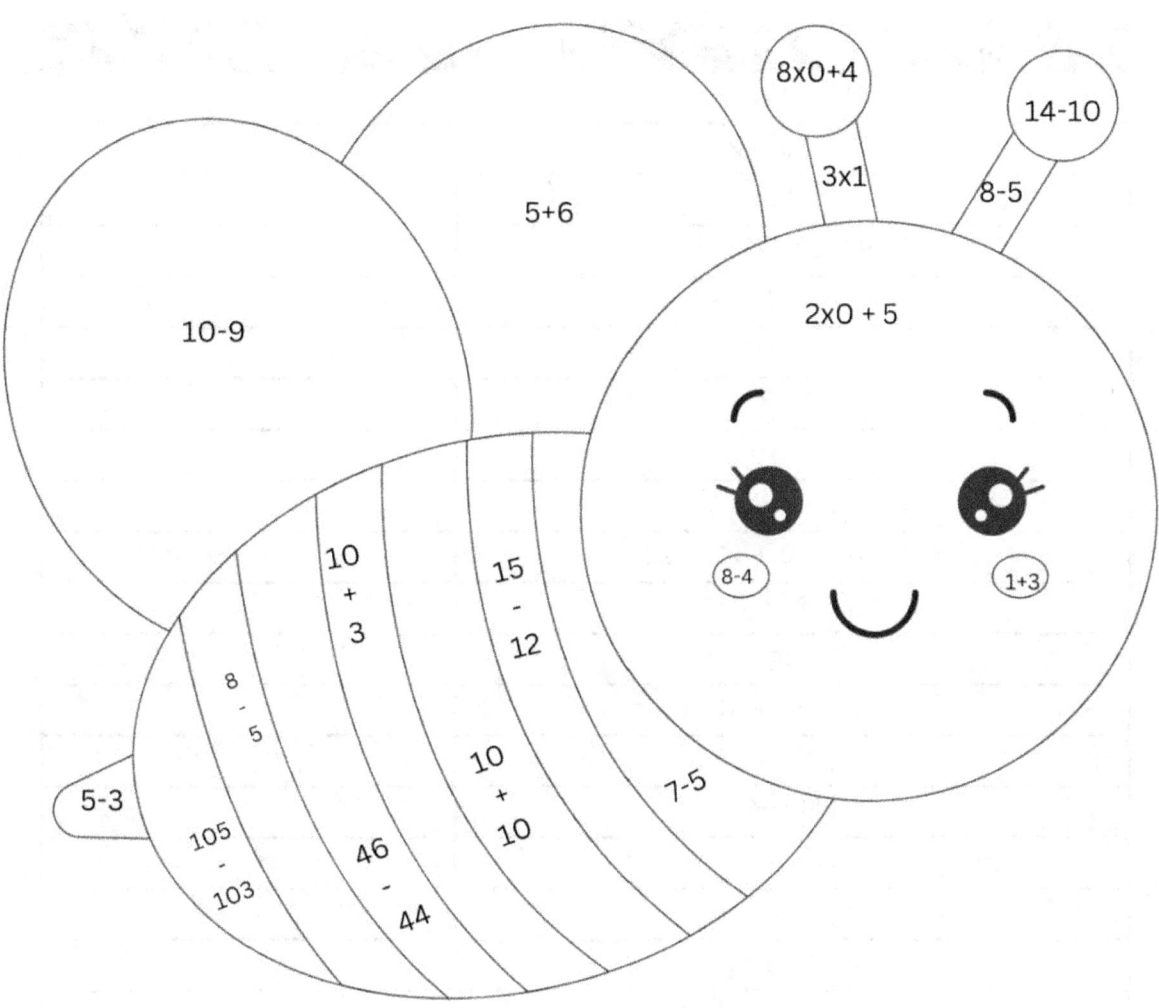

PUZZLE 14

Solve the equations and connect the dots in order

MATHS PROBLEM	ANSWER
Start: 100 - 10	
27 + 20 + 20	
20 +15	
90, 80, 70, ?	
200 + 100 +200 + 100	
Half of 12	
30 - 6	
50 - 10	
3 + 18	
20 + 20 + 20 + 20 + 20	
52, 62, ?, 82	
1/2 of 6	
26 + 10 + 10	
? , 38, 58, 78	
1/4 of 4	
2 x 4	
41 + 10 + 10	
3 x 3	
10 + 10 + 5	
9 x ? = 90	
24 + 20	
25 + 30	
3 + 4	
5, 10, ? 20	
80 + 7	
100 + 100	
15 + 30	
8, 12, ? 20	

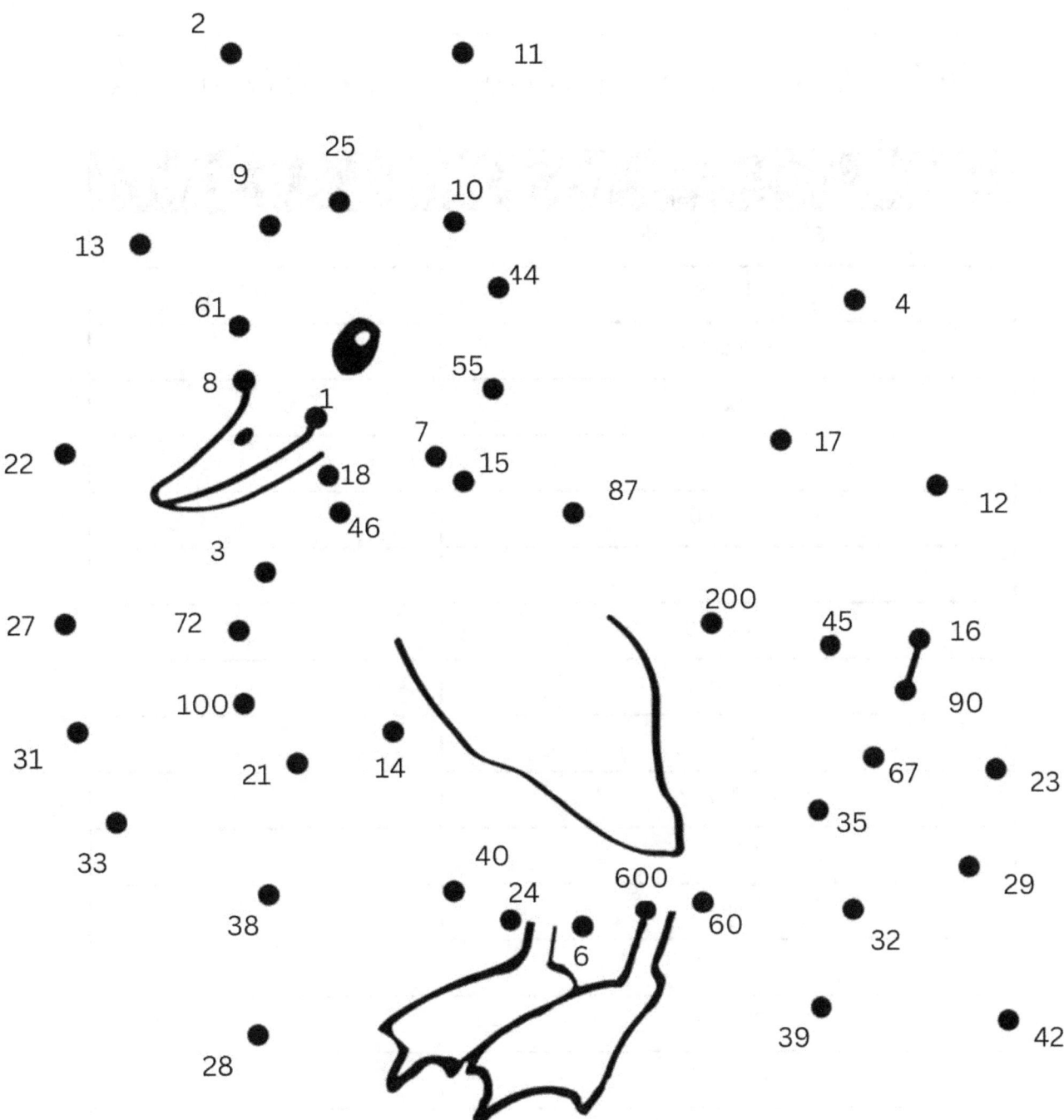

PUZZLE 15

Solve the equations and connect the dots in order

MATHS PROBLEM	ANSWER
Start: Half of 100	
70 + 8	
Half of 50	
4 x 4	
31, ?, 21, 16	
23 + 60	
100 -1	
3 x ? = 9	
4 + ? = 40	
10 - 3	
15, ?, 7, 3	
50 - 7	
70 + 20	
77 - 11	
20 - 12	
152 - 52	
28 + 5	
7, 14, ?, 28	
12 + 5	
?, 7, 12, 17	

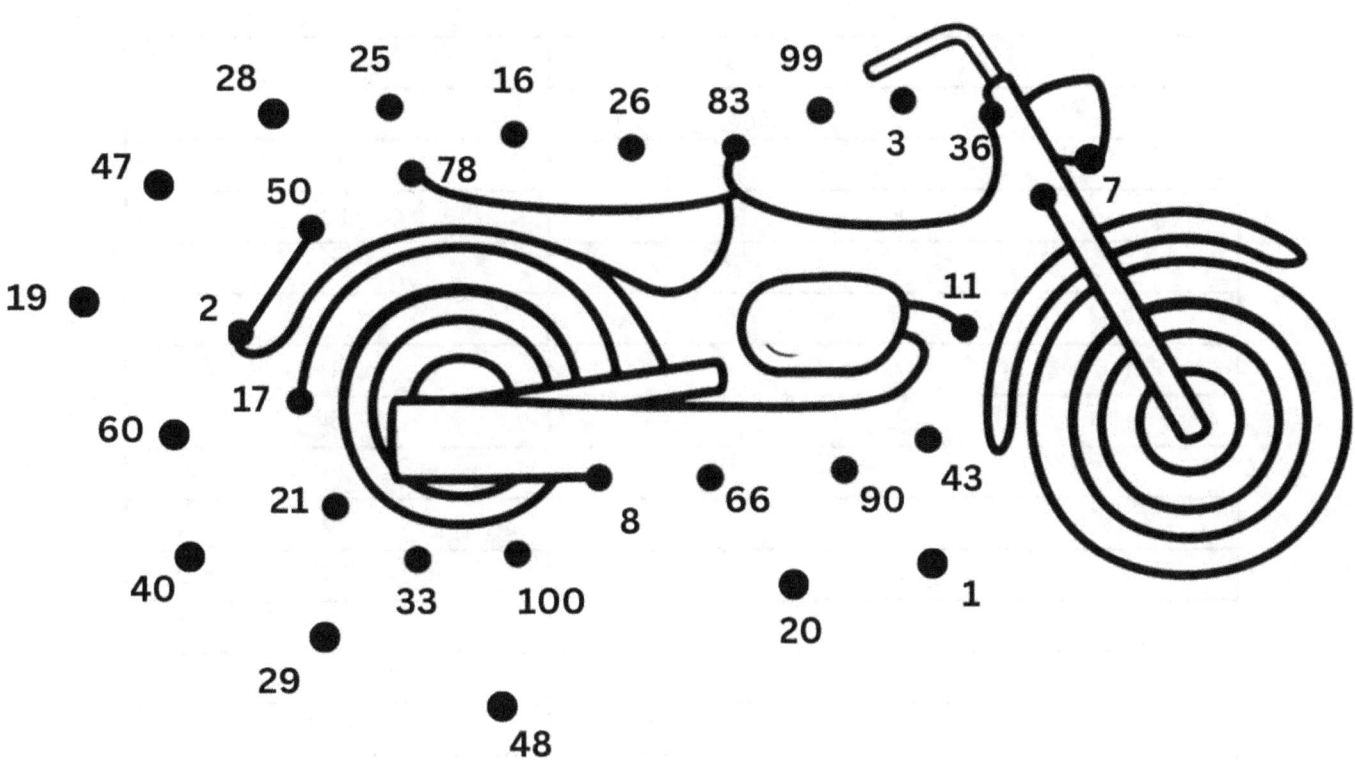

PUZZLE 16

Solve the equations to find the letters

LETTER	ANSWER
A	3
B	21
C	200
D	100
E	2
F	50
G	5
H	12
I	29
J	56
K	4
L	6
M	25
N	20
O	14
P	150
Q	400
R	1000
S	10
T	7
U	46
V	36
W	110
X	240
Y	9
Z	68

What time does a dentist eat lunch?

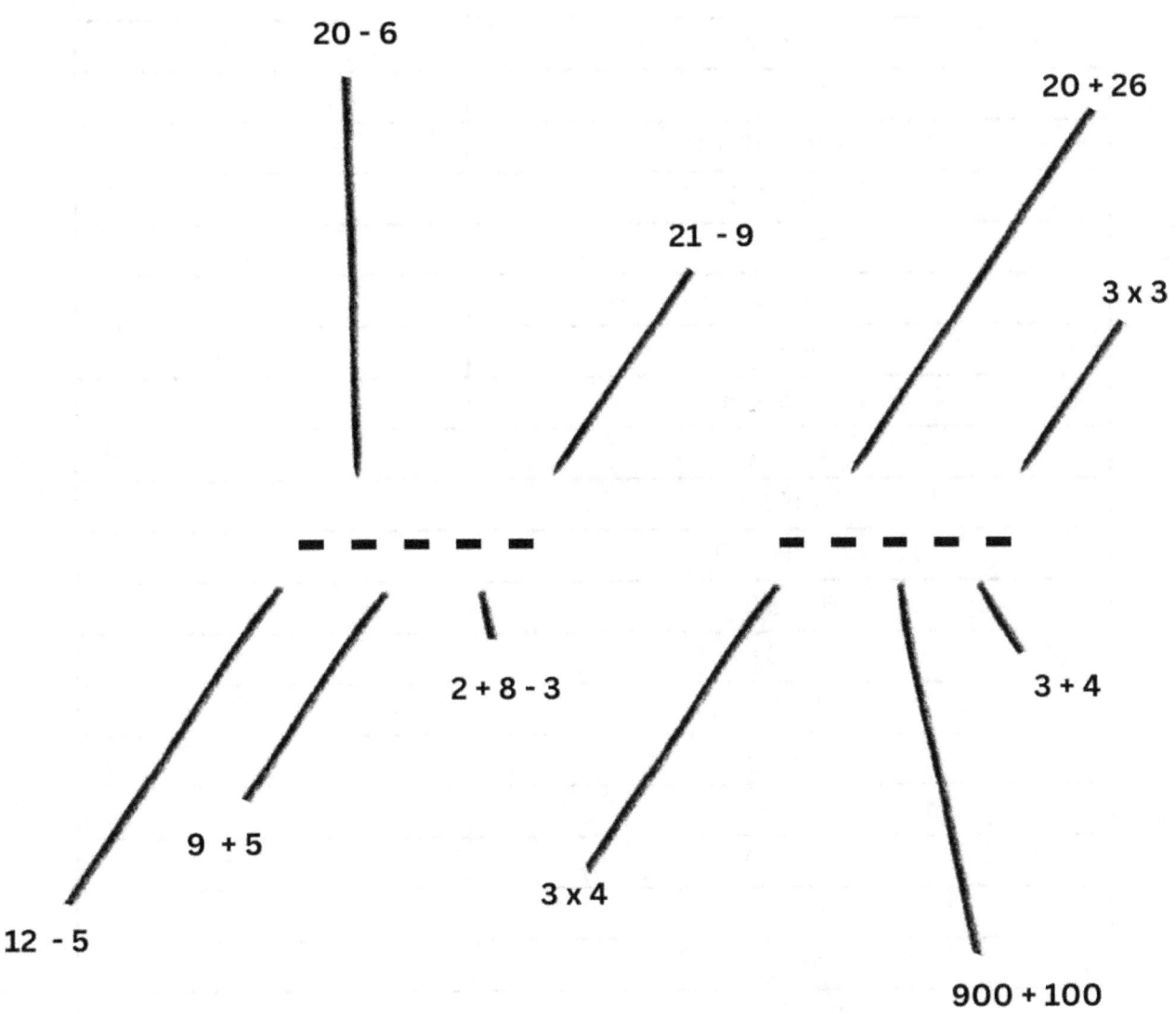

PUZZLE 17

Solve the equations and connect the dots in order

MATHS PROBLEM	ANSWER
Start: 90 + 10	
96, 86, ?, 66	
128 - 100	
40 -3	
20 + 30 +2	
100 - 9	
10 + 10 + 10 + 9	
70, 65, ?, 55	
40 - 31	
91 + 8	
30 - 7	
10 + 15 + 20	
15 - 14	
Half of 100	
12 + 20 + 10	
Half of 6	
10 + 20 - 4	
4 x 10	
? x 4 =20	
40 - 8	
100 - 30	
36 + 8	
26 + 8	
60 + 6 - 10	
28, 21, 14, ?	
40 + 40 + 8	
300 -200	

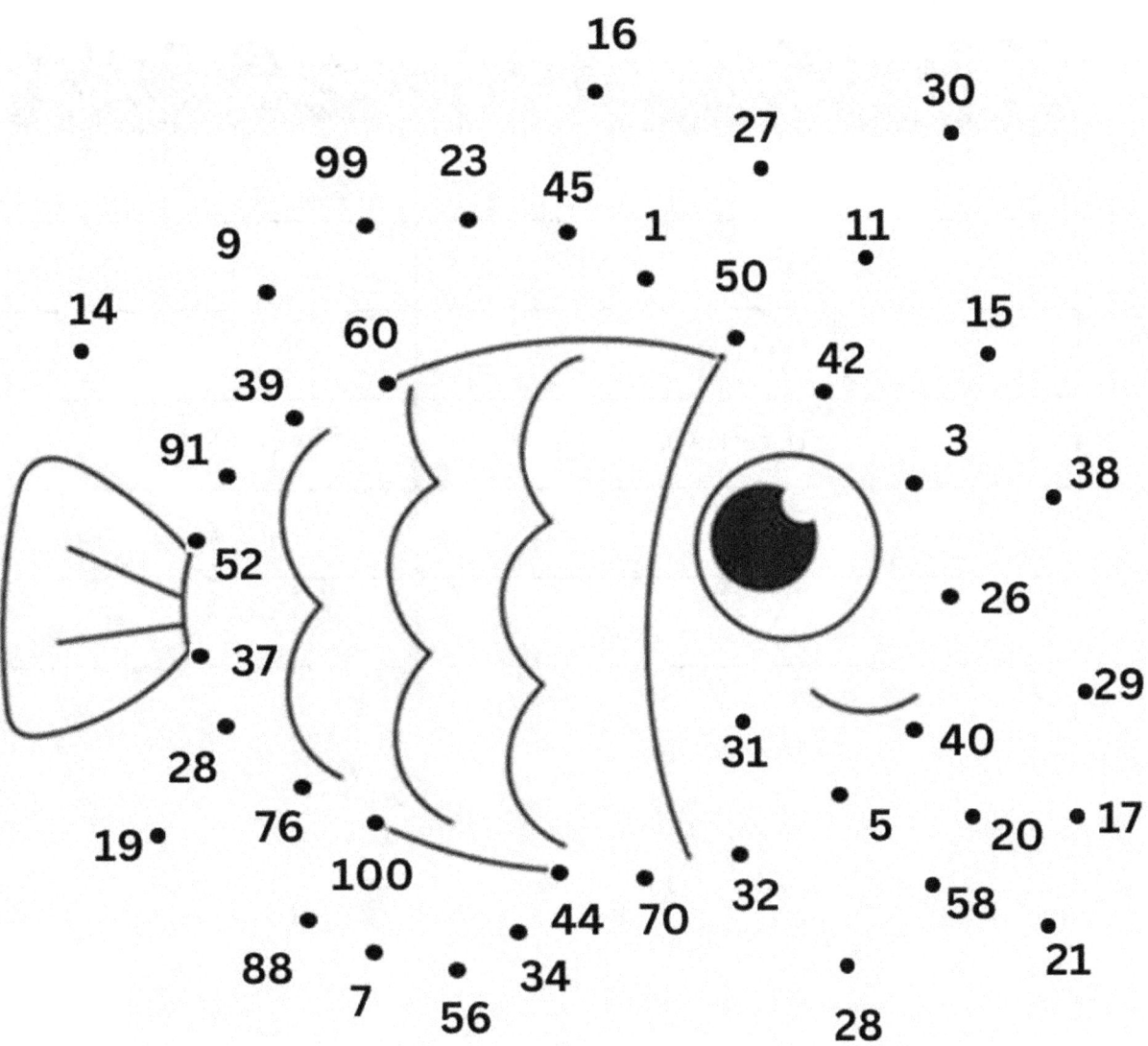

PUZZLE 18

Colour By Number

MATHS PROBLEM	ANSWER
Red	2
Orange	3
Grey	4
Light blue	5
Black	6
Yelow	7

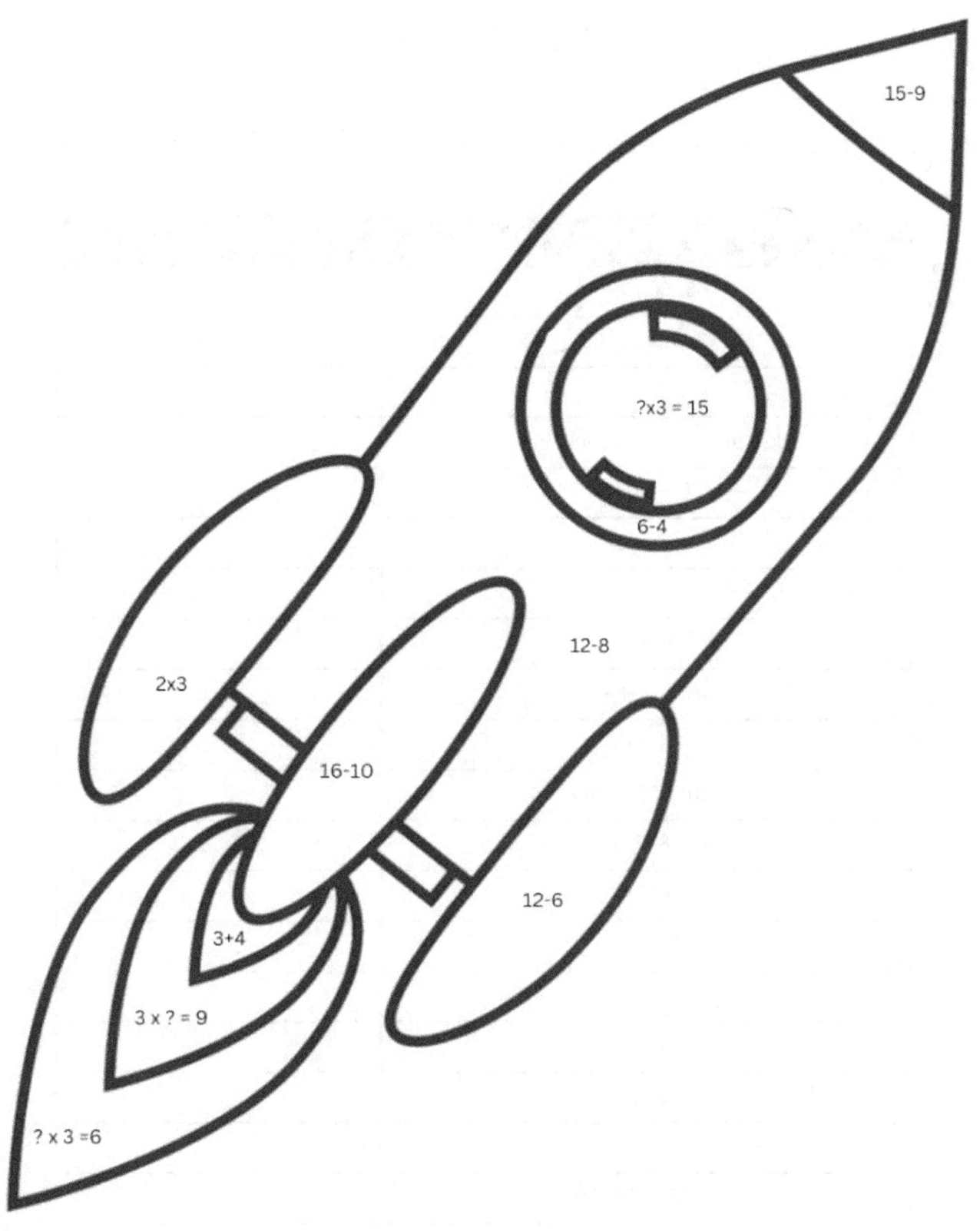

PUZZLE 19

Solve the equations and connect the dots in order

MATHS PROBLEM	ANSWER
55, 66, ?, 88	
Quarter of 100	
10 - 2	
How many days in a fortnight?	
20 + 40 +6	
9 - 6	
40 - 1	
100 -9	
4 + 4 + 4 + 4	
34 +10	
20 + 50 + 10	
4, 8, ?, 16	
70 - 4	
10 +12	
Half of 12	
2 x ? =8	
2 x 5	
20 -1	
50, ?, 150, 200	
a quarter (1/4) of 8	

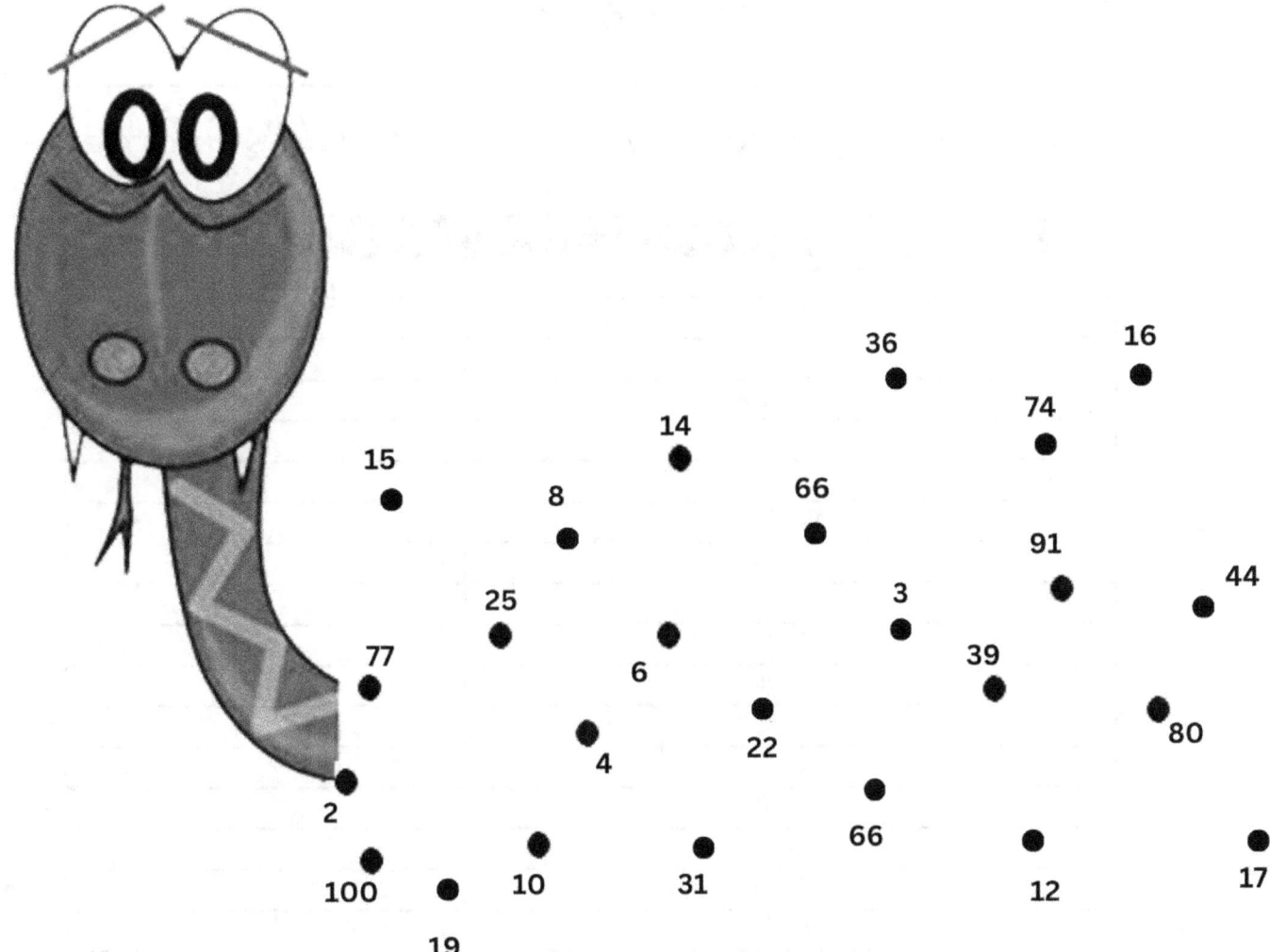

PUZZLE 20

Solve the equations and connect the dots in order

MATHS PROBLEM	ANSWER
A quarter (1/4) of 4	
20 - 3	
2 x 2	
20, 40, ?, 80	
3 x 10	
3 x 3	
60 - 5	
12 + 10	
16 + 12	
10 + 9	
42 + 20	
15, 13, ?, 9	
40 - 7	
100 - 8	
6 + 7	
Half of 40	
50 + 50	
4 x 4	
55, 61, ?, 71	
Half of 80	
50 + 27	
40 - 3	
9 x 10	
? x 4 = 12	
10 +190	
? + 25 = 50	
80 + 4	
22, 18, ?, 10	
50 - 5	
12 - 4	
How many minutes in a quarter of an hour?	
42 - 4	
106, 96, 86, ?	

PUZZLE 21

Solve the equations and connect the dots in order

MATHS PROBLEM	ANSWER
Start: 20 - 4	
Half of 44	
? x 5 = 10	
10 + 2	
? x 3 = 15	
50 - 13	
10, 15, 20, ?	
2 x ? = 8	
60 - 2	
50 + 16	
80 - 3	
50 + 11	
? + 7 = 60	
20 + 16	
19, 14, ?, 4	
41 + 2	
10 + 10 + 10 + 8	
91 - 3	
10 + 10 + 4	
? x 2 = 6	
? - 2 = 50	
11, ?, 17, 20	
19 + 4	
32 + 7	
1/2 of 2	
50 - 4	

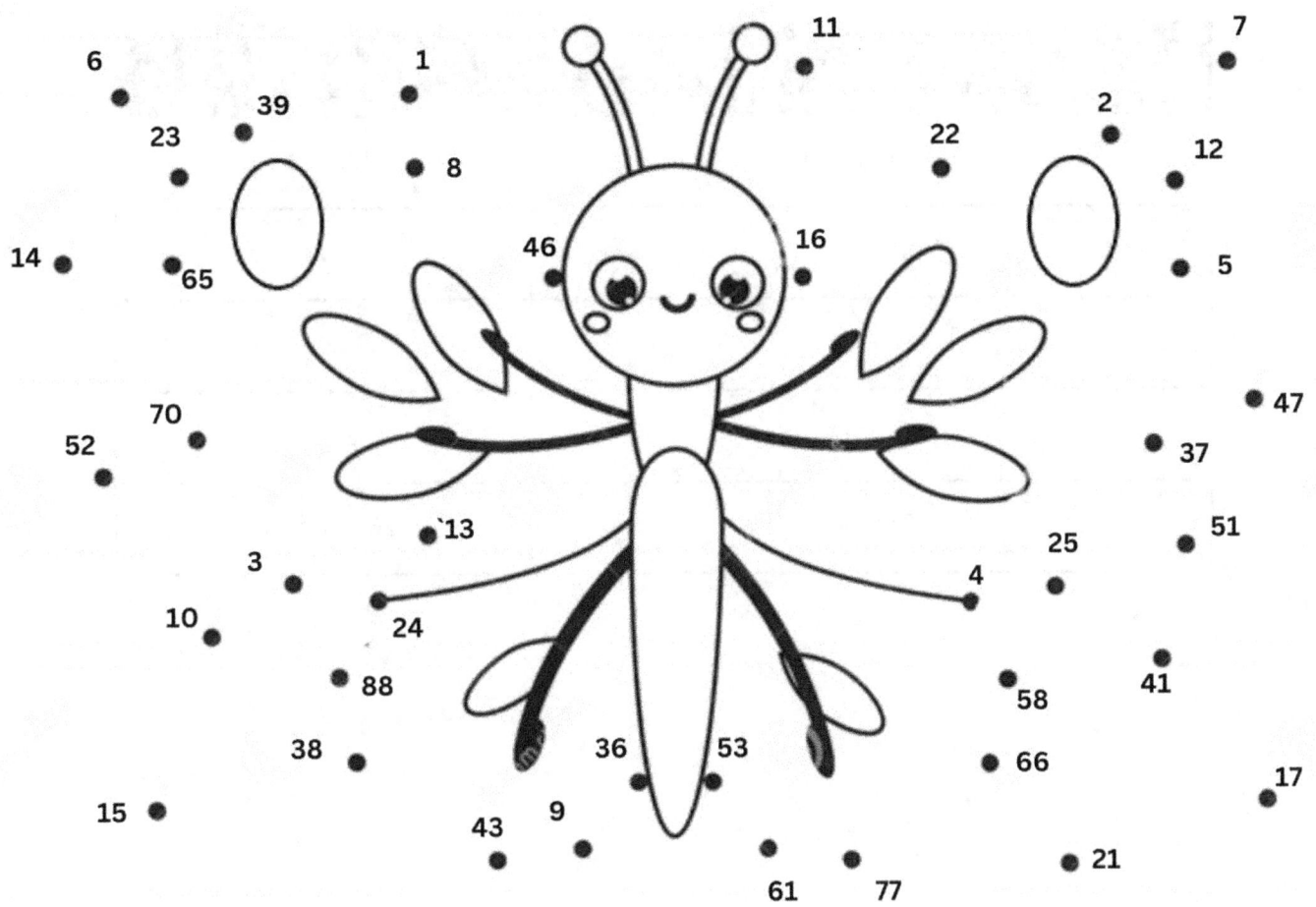

PUZZLE 22

Colour By Number

MATHS PROBLEM	ANSWER
Light Green	2,5 or 12
Dark Green	4,8,60 or 76
Red	9,33 or 100
Brown	6, 23 or 45
Orange	26, 34 or 52

PUZZLE 23

Solve the equations to find the letters

LETTER	ANSWER
A	3
B	21
C	200
D	100
E	2
F	50
G	5
H	12
I	29
J	56
K	4
L	6
M	25
N	20
O	14
P	150
Q	400
R	1000
S	10
T	7
U	46
V	36
W	110
X	240
Y	9
Z	68

What do bees use to brush their hair?

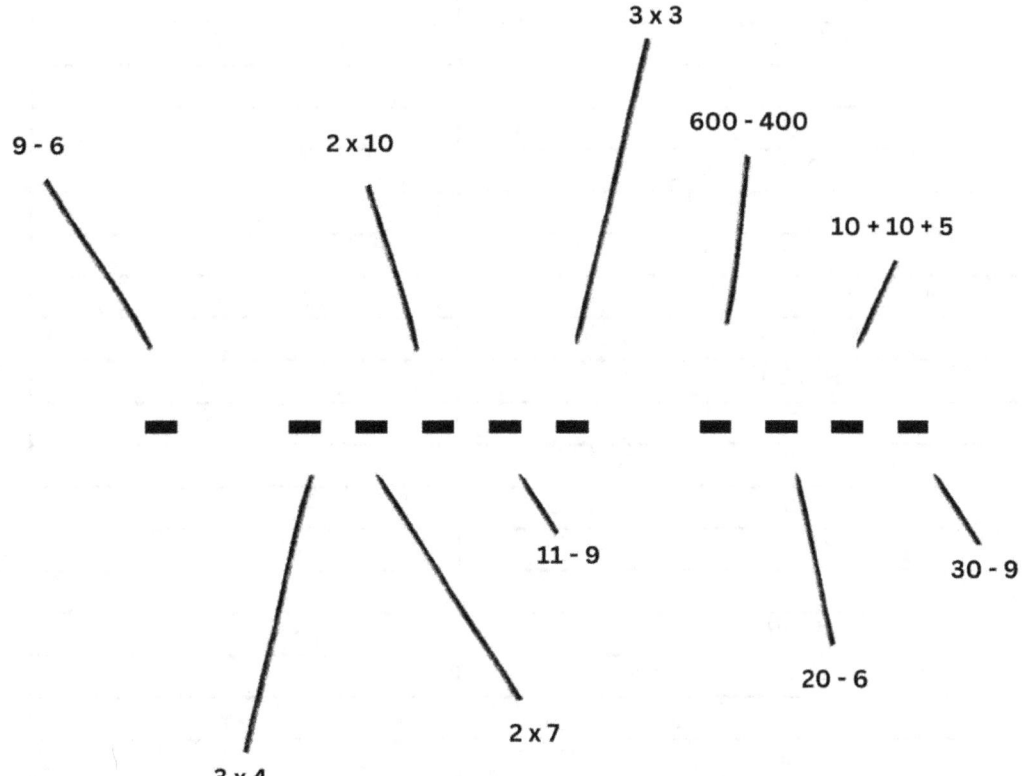

PUZZLE 24

Solve the equations and connect the dots in order

MATHS PROBLEM	ANSWER
Start: A quarter of 4	
55, 44, ?, 22	
10 x 2	
3 x 3	
2 x 2	
50 + 11	
20 + 20 +15	
Half of 22	
28, 21, 14, ?	
20 +20 +20 +5	
30 - 6	
2 x 4	
100 - 1	
4 x 10	
60 - 8	
3 x 4	
45, 35, ?, 15	
Half of 200	
2 x 3	
30 - 3	
2 x 5	
37 + 10	
20 + 20 + 18	
Half of 10	
How many minutes in a quarter of an hour?	
300 - 100	
7 + 6	
10 + 10 +9	
80 - 8	
Half (1/2) of 6	
11, ?, 33, 44	
20 + 30	
7 - 5	
25 -4	

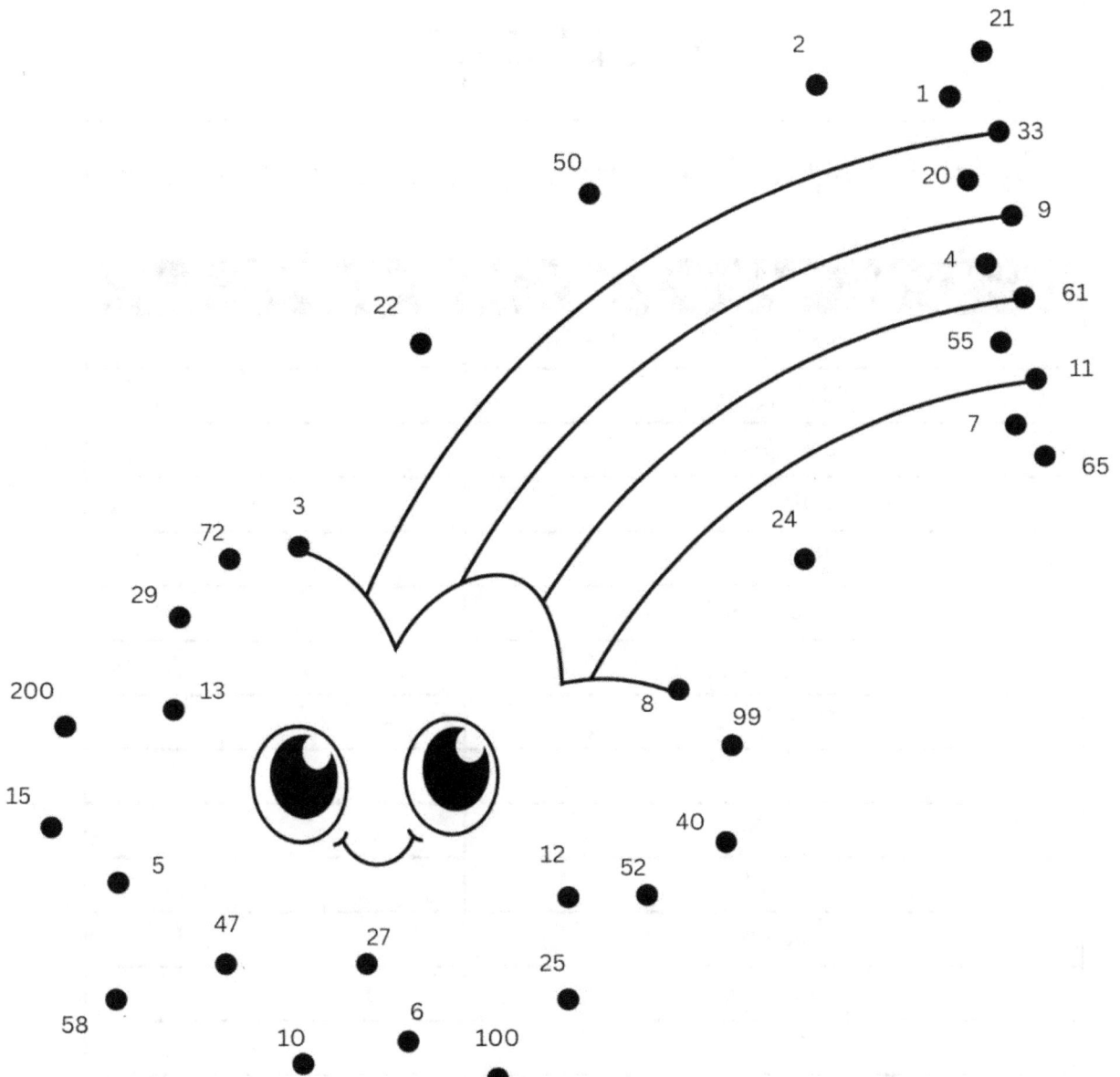

PUZZLE 25

Solve the equations and connect the dots in order

MATHS PROBLEM	ANSWER
Start: 47 -46	
11, 22, ?, 44	
20 + 20 + 20	
10 + 7	
800 - 700	
28, 21, 14, ?	
50 - 6	
10 + 6 + 3	
7 - 5	
3 x 3	
2 x 11	
60 - 6	
22 + 10	
90 - 4	
Half of 8	
2 x 5	
24, 18, 12, ?	
60 - 5	
3 x 4	
11 + 10	
10 + 10 + 10 + 8	
20 + 20 + 6	
7, ?, 19, 25	

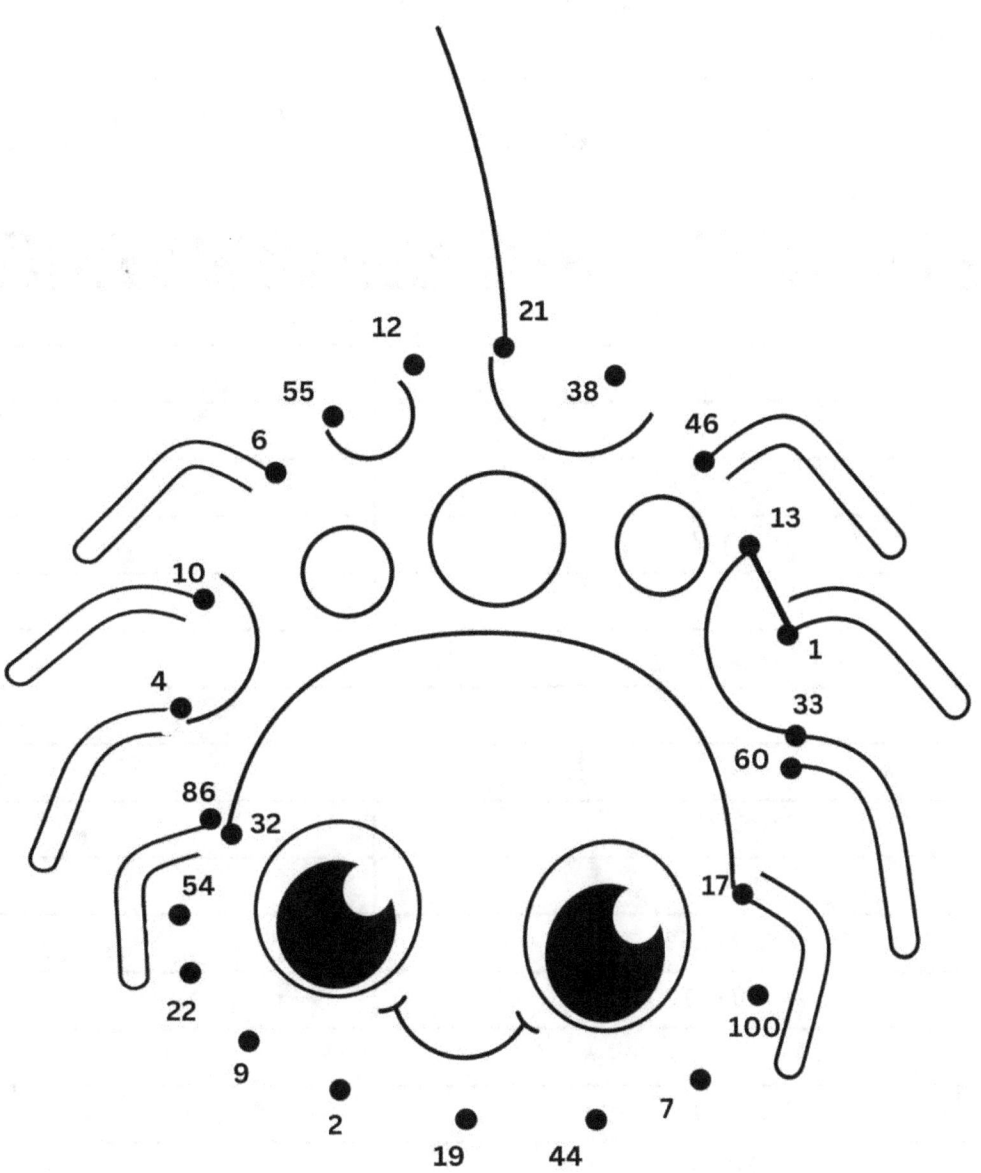

PUZZLE 26

Solve the equations and connect the dots in order

MATHS PROBLEM	ANSWER
Start: 16 + 10	
10 - 5	
126 -26	
23 + 10	
31, 34, ?, 40	
33 -10	
20 + 30	
3 + 10	
28 - 27	
71, 51, 31, ?	
half of 4	
Half of 44	
10 + 10 + 10 + 6	
20 +20 + 10 + 9	
28, 21, 14, ?	
50 - 9	
70 + 7	
3 x 3	
50 - 2	
70 + 11	
?, 9, 15, 21	
Half of 20	
20 + 20 + 20 + 20 + 8	
50 + 23	

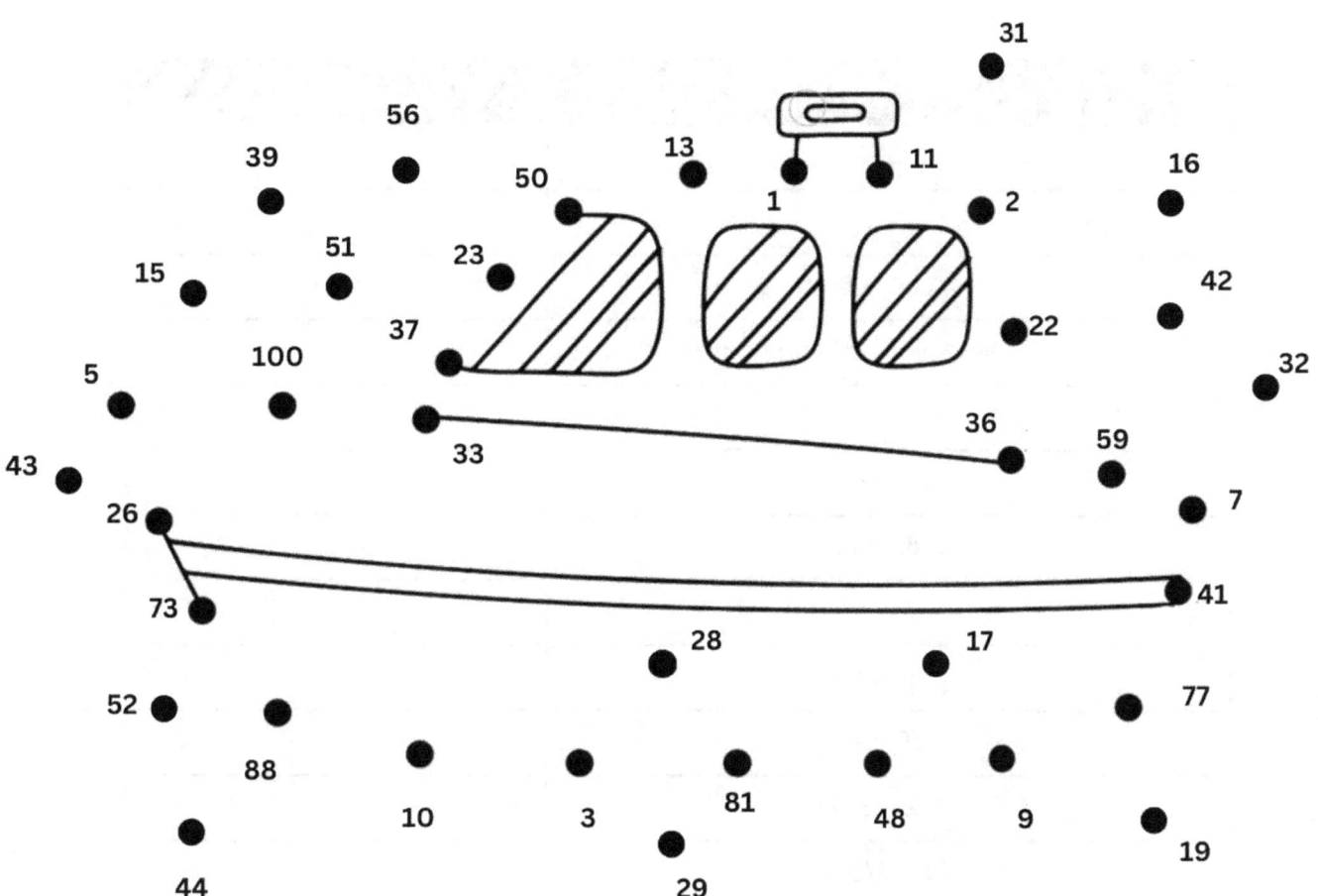

PUZZLE 27

Solve the equations and connect the dots in order

MATHS PROBLEM	ANSWER
Start: 3+2 - 4	
42, 45, ?, 51	
11 + 11	
How many minutes in an hour?	
3 x 3	
29 + 4	
78, 82, 86, ?	
2 x 2	
Half of 80	
50 - 5	
27, 25, ?, 21	
a third (1/3) of 9	
7 x 10	
79, 59, 39, ?	
50 + 21	
1/2 of 4	
20 + 20 + 20 + 3	
10 + 20 + 25	
40 - 11	
A quarter (1/4) of 4	

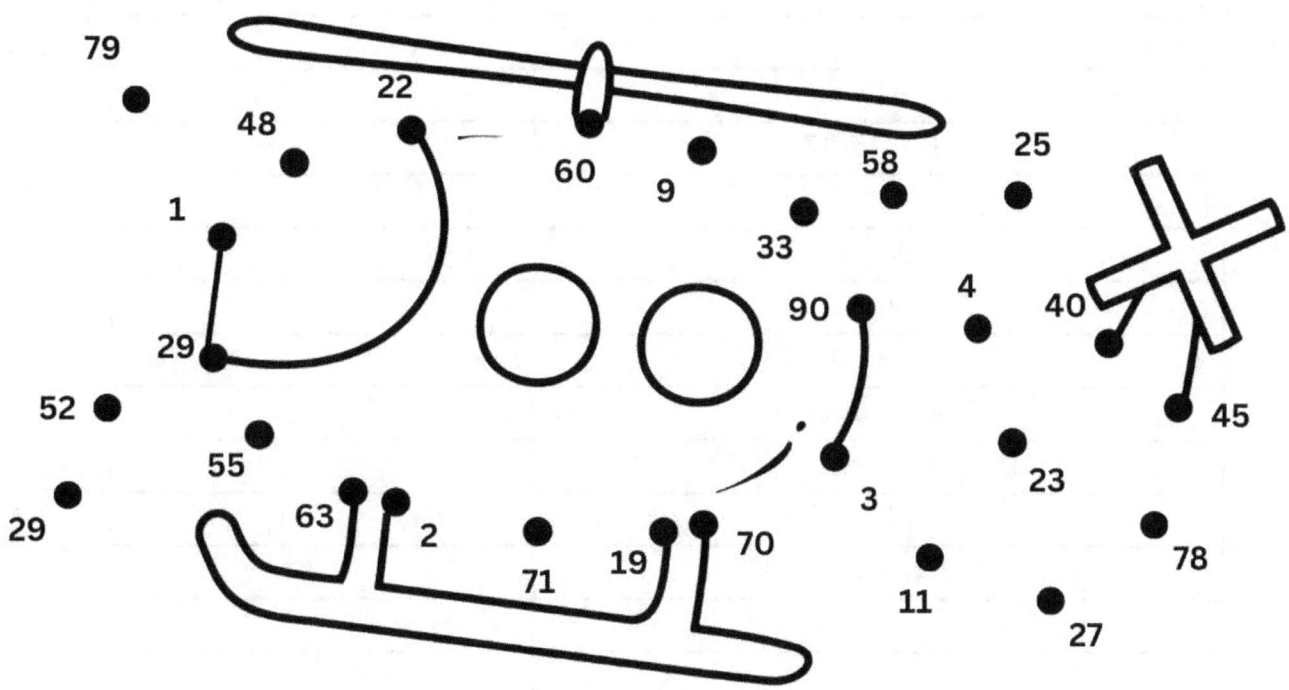

PUZZLE 28

Solve the equations and connect the dots in order

MATHS PROBLEM	ANSWER
Start: 50 - 10 - 3	
2 x 4	
90 - 8	
10 + 6	
2 x 3	
14 + 10 + 4	
11, ?, 15, 17	
4 x 10	
Half of 6	
100 - 9	
11 + 10	
28, 21, 14, ?	
21 - 10	
10 - 8	
16 + 10	
50 + 10 + 10	
8 - 7	
10 + 8	
70, 50, 30, ?	
20 + 24	
20 + 30 + 38	
3 x 3	
22, 44, ?, 88	
11 x 5	
10 x 10	
80 - 4	
Half of 40	

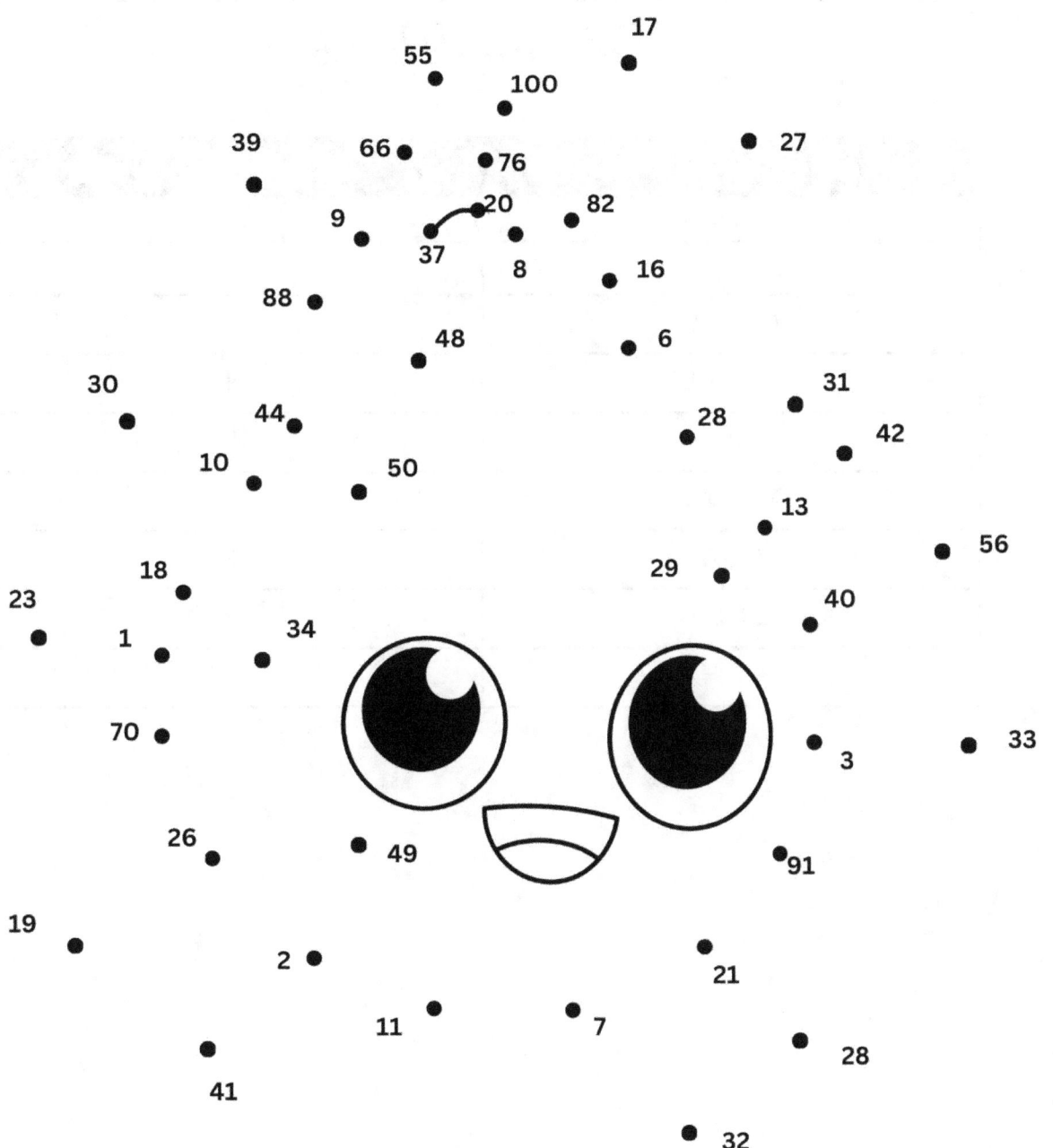

PUZZLE 29

Colour by Number

MATHS PROBLEM	ANSWER
RED	1
GREY	2
ORANGE	3
LIGHT BLUE	4
BLACK	5
YELLOW	6
WHITE	7
GREEN	8
PINK	9

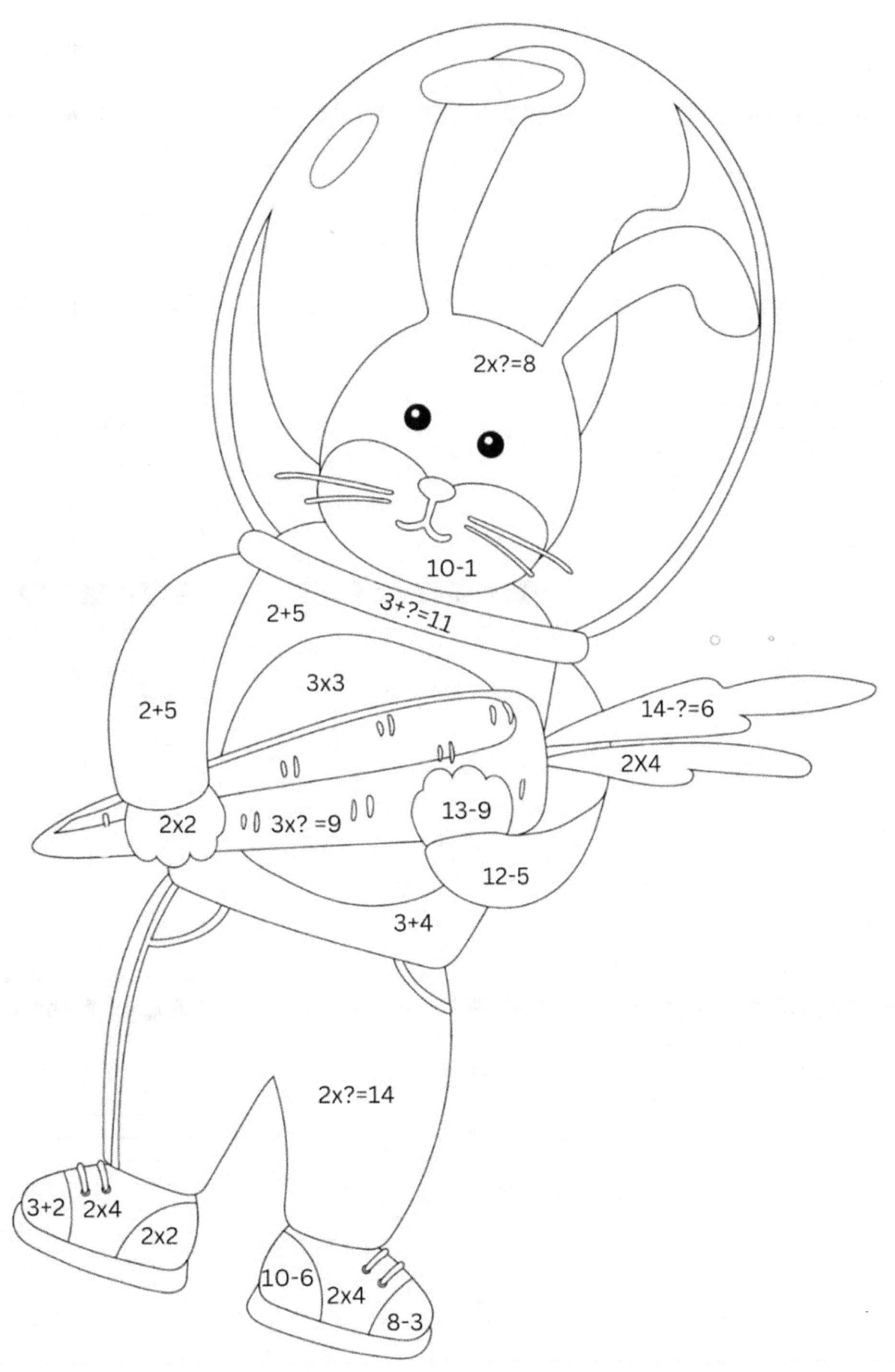

ANSWERS

PUZZLE 1
Solve the equations and connect the dots in order

MATHS PROBLEM	ANSWER
Start: 3 - 2	1
10 + 7	17
32 - 4	28
10 - 7	3
7, 9, ?, 13	11
10 x 10	100
half of 4	2
7 + ? = 20	13
6 + ? =20	14
10 + 10 + 6	26
3 x 10	30
4,7,10,13	7
2 x 5	10
30 - 7	23

PUZZLE 2
Solve the equations and connect the dots in order

MATHS PROBLEM	ANSWER
3 + ? = 10	7
3,6,?,12	9
5 + 6	11
2 x 10	20
20 +20 + 2	42
4, 8, 12, ?	16
70 - 3	67
10 + 20 + 6	36
Half of 100	50

PUZZLE 3
Solve the equations and connect the dots in order

MATHS PROBLEM	ANSWER
START: 12 - 16	1
6 + 7	13
30 + 30 + 30	90
100, ?, 300, 400	200
30 + 40 + 2	72
2 x 4	8
10 + ? = 13	3
2 x 40	80
37, 39, ?, 43	41
50 - 6	44
30 + 30	60
50 + 5	55
half of 8	4
10-8	2
100 - 12	88
11, 22, ?, 44	33
20 +20 + 10 +7	57
30 - 9	21
Half of 10	5
30 + 30 + 30 + 1	91
How many minutes in a quarter of an hour?	15
10 x 10	100
8 + 8	16
6+5	11
30 - 10 - 19	1

PUZZLE 4
Solve the equations the find the letters

You Planet

PUZZLE 5
Solve the equations and connect the dots in order

MATHS PROBLEM	ANSWER
20 + ? = 70	50
90, 100, ?, 120	110
11 x 2	22
How many minutes in an hour?	60
100 - 99	1
50 + 50 +100	200
10 + 10 + 20	40
99 + 24	123
101, 100, ?, 98	99
100 - 10	90
10 - 3	7
7 + 8	15

PUZZLE 6
Solve the equations and connect the dots in order

MATHS PROBLEM	ANSWER
10 +10 +10 + 7	37
20 x 3	60
10 +10 +7	27
Half of 10	5
53, 56, ?, 62	59
15 - 5 - 9	1
20 x 2	40
6, ?, 26, 36	16
40 -14	26
1 + 9 +40	50

PUZZLE 7
Colour by Numbers

MATHS PROBLEM	ANSWER
BLACK	11
RED	5
GREY	3
LIGHT BLUE	4
YELLOW	2
ORANGE	6

PUZZLE 8
Solve the equations and connect the dots in order

MATHS PROBLEM	ANSWER
13 - 12	1
2 x 4	8
8, ?, 4, 2	6
100 + 200 + 100	400
How many days in a fortnight?	14
6 x ? = 12	2
50 - 9	41
80 - 3	77
17 + 6	23
7, 9, ?, 13	11
11 x 2	22
3 x 3	9
80 -14	66
15, ?, 9, 6	12
? - 16 = 10	26
20 + 30 + 4	54
50 + 20 + 30	100
3, 9, ?, 21	15
5 x ? = 15	3
90 -9	81
50 + 10 + 1	61
2000 -1000	1000
23 +10	33
1, ?, 9, 13	5
30 + 30 + 30 + 1	91
50 + 50 + 6	106
Half of 20	10
3 x 5	15

PUZZLE 9
Solve the equations and connect the dots in order

MATHS PROBLEM	ANSWER
1/4 of 4	1
126 - 100	26
How many days in a fortnight?	14
51 + 20	71
20 - 10	10
3 + 4	7
19 - 8	11
6 x 10	60
Half of 10	5
40 - 20	20
16 + ? = 20	4
10 + 10 + 10 + 6	36
10 x 10	100

PUZZLE 10

Solve the equations and connect the dots in order

MATHS PROBLEM	ANSWER
START: 8 - 5 - 2	1
101 - 2	99
3 x 4	12
3 x 3	9
10 - 3	7
31 + 10 + 10	51
20 - 10	10
21 - 2	19
27, 37, ?, 57	47
68 + 20	88
23 = 22 + ?	1

PUZZLE 11

Solve the equations the find the letters

A Wooly Jumper

PUZZLE 12

Colour By Number

MATHS PROBLEM	ANSWER
BLACK	10
RED	6
DARK BLUE	1
ORANGE	7
SILVER	9
GREY	5
WHITE	3
YELLOW	2
PURPLE	12

PUZZLE 13

Colour By Number

MATHS PROBLEM	ANSWER
light blue	1, 8 or 11
yellow	3, 13 or 100
black	2, 20 or 34
orange	5, 9 or 87
pink	4, 51 or 68

PUZZLE 14

Solve the equations and connect the dots in order

MATHS PROBLEM	ANSWER
Start: 100 - 10	90
27 + 20 + 20	67
20 + 15	35
90, 80, 70, ?	60
200 + 100 + 200 + 100	600
Half of 12	6
30 - 6	24
50 - 10	40
3 + 18	21
20 + 20 + 20 + 20 + 20	100
52, 62, ?, 82	72
1/2 of 6	3
26 + 10 + 10	46
?, 38, 58, 78	18
1/4 of 4	1
2 x 4	8
41 + 10 + 10	61
3 x 3	9
10 + 10 + 5	25
9 x ? = 90	10
24 + 20	44
25 + 30	55
3 + 4	7
5, 10, ? 20	15
80 + 7	87
100 + 100	200
15 + 30	45
8, 12, ? 20	16

PUZZLE 15

Solve the equations and connect the dots in order

MATHS PROBLEM	ANSWER
Start: Half of 100	50
70 + 8	78
Half of 50	25
4 x 4	16
31, ?, 21, 16	26
23 + 60	83
100 - 1	99
3 x ? = 9	3
4 + ? = 40	36
10 - 3	7
15, ?, 7, 3	11
50 - 7	43
70 + 20	90
77 - 11	66
20 - 12	8
152 - 52	100
28 + 5	33
7, 14, ?, 28	21
12 + 5	17
?, 7, 12, 17	2

PUZZLE 16

Solve the equations to find the letters

Tooth hurty

PUZZLE 17

Solve the equations and connect the dots in order

MATHS PROBLEM	ANSWER
Start: 90 + 10	100
96, 86, ?, 66	76
128 - 100	28
40 - 3	37
20 + 30 + 2	52
100 - 9	91
10 + 10 + 10 + 9	39
70, 65, ?, 55	60
40 - 31	9
91 + 8	99
30 - 7	23
10 + 15 + 20	45
15 - 14	1
Half of 100	50
12 + 20 + 10	42
Half of 6	3
10 + 20 - 4	26
4 x 10	40
? x 4 = 20	5
40 - 8	32
100 - 30	70
36 + 8	44
26 + 8	34
60 + 6 - 10	56
28, 21, 14, ?	7
40 + 40 + 8	88
300 - 200	100

PUZZLE 18

Colour By Number

MATHS PROBLEM	ANSWER
Red	2
Orange	3
Grey	4
Light blue	5
Black	6
Yellow	7

PUZZLE 19

Solve the equations and connect the dots in order

MATHS PROBLEM	ANSWER
55, 66, ?, 88	77
Quarter of 100	25
10 - 2	8
How many days in a fortnight?	14
20 + 40 + 6	66
9 - 6	3
40 - 1	39
100 - 9	91
4 + 4 + 4 + 4	16
34 + 10	44
20 + 50 + 10	80
4, 8, ?, 16	12
70 - 4	66
10 + 12	22
Half of 12	6
2 x ? = 8	4
2 x 5	10
20 - 1	19
50, ?, 150, 200	100
a quarter (1/4) of 8	2

PUZZLE 20

Solve the equations and connect the dots in order

MATHS PROBLEM	ANSWER
A quarter (1/4) of 4	1
20 - 3	17
2 x 2	4
20, 40, ?, 80	60
3 x 10	30
3 + 3	9
60 - 5	55
17 + 10	27
16 + 12	28
10 + 9	19
42 + 20	62
15, 13, ?, 9	11
40 - 7	33
100 - 8	92
6 + 7	13
Half of 40	20
50 + 50	100
4 + 4	8
55, 61, ?, 71	66
Half of 80	40
50 + 27	77
40 - 3	37
9 x 10	90
? + 4 = 12	1
10 + 90	100
? + 25 = 50	25
80 + 4	84
22, 18, ?, 10	14
50 - 5	45
12 - 4	8
How many minutes in a quarter of an hour?	15
42 - 4	38
106, 96, 86, ?	76

PUZZLE 21

Solve the equations and connect the dots in order

MATHS PROBLEM	ANSWER
Start: 20 - 4	16
Half of 44	22
? x 5 = 10	2
10 + 2	12
? x 3 = 15	5
50 - 13	37
10, 15, 20, ?	25
2 x ? = 8	4
60 - 2	58
50 + 16	66
80 - 3	77
50 + 11	61
? + 7 = 60	53
20 + 16	36
19, 14, ?, 4	9
41 + 2	43
10 + 10 + 10 + 8	38
91 - 3	88
10 + 10 + 4	24
? x 2 = 6	3
? - 2 = 50	52
11, ?, 17, 20	14
19 + 4	23
32 ÷ ?	39
1/2 of 2	1
50 - 4	46

PUZZLE 22

Colour By Number

MATHS PROBLEM	ANSWER
Light Green	2, 5 or 12
Dark Green	4, 8, 60 or 76
Red	9, 33 or 100
Brown	6, 23 or 45
Orange	26, 34 or 52

PUZZLE 23

Solve the equations to find the letters

A honey comb

PUZZLE 24

Solve the equations and connect the dots in order

MATHS PROBLEM	ANSWER
Start: A quarter of 4	1
55, 44, ?, 22	33
18 + 2	20
3 x 3	9
2 x 2	4
50 + 11	61
20 + 20 + 15	55
Half of 22	11
28, 21, 14, ?	7
20 + 20 + 20 + 5	65
30 - 6	24
2 x 4	8
100 - 1	99
4 x 10	40
60 - 8	52
3 x 4	12
45, 35, ?, 15	25
Half of 200	100
2 x 3	6
30 - 3	27
2 x 5	10
37 + 10	47
20 + 20 + 18	58
Half of 10	5
How many minutes in a quarter of an hour?	15
300 - 100	200
7 + 6	13
10 + 10 + 9	29
80 - 8	72
Half (1/2) of 6	3
11, ?, 33, 44	22
70 - 30	50
7 - 5	2
25 - 4	21

PUZZLE 25

Solve the equations and connect the dots in order

MATHS PROBLEM	ANSWER
Start: 47 - 46	1
11, 22, ?, 44	33
20 + 20 + 20	60
10 + 7	17
800 - 700	100
28, 21, 14, ?	7
50 - 6	44
10 + 6 + 3	19
7 - 5	2
3 x 3	9
2 x 11	22
60 - 6	54
22 + 10	32
90 - 4	86
Half of 8	4
2 x 5	10
24, 18, 12, ?	6
60 - 5	55
3 x 4	12
11 + 10	21
10 + 10 + 10 + 8	38
20 + 20 + 6	46
7, ?, 19, 25	13

PUZZLE 26

Solve the equations and connect the dots in order

MATHS PROBLEM	ANSWER
Start: 16 + 10	26
10 - 5	5
126 - 26	100
23 + 10	33
31, 34, ?, 40	37
33 - 10	23
20 + 30	50
3 + 10	13
28 - 27	1
71, 51, 31, ?	11
half of 4	2
Half of 44	22
10 + 10 + 10 + 6	36
20 + 20 + 10 + 9	59
28, 21, 14, ?	7
50 - 9	41
70 + 7	77
3 x 3	9
50 - 2	48
70 + 11	81
?, 9, 15, 21	3
Half of 20	10
20 + 20 + 20 + 20 + 8	88
50 + 23	73

PUZZLE 27

Solve the equations and connect the dots in order

MATHS PROBLEM	ANSWER
Start: 3 + 2 - 4	1
42, 45, ?, 51	48
11 + 11	22
How many minutes in an hour?	60
3 x 3	9
29 + 4	33
78, 82, 86, ?	90
2 x 2	4
Half of 80	40
50 - 5	45
27, 25, ?, 21	23
a third (1/3) of 9	3
7 x 10	70
79, 59, 39, ?	19
50 + 21	71
1/2 of 4	2
20 + 20 + 20 + 3	63
10 + 20 + 25	55
40 - 11	29
A quarter (1/4) of 4	1

PUZZLE 28

Solve the equations and connect the dots in order

MATHS PROBLEM	ANSWER
Start: 50 - 10 - 3	37
2 x 4	8
90 - 8	82
10 + 6	16
2 x 3	6
14 + 10 + 4	28
11, ?, 15, 17	13
4 x 10	40
Half of 6	3
100 - 9	91
11 + 10	21
28, 21, 14, ?	7
21 - 10	11
10 - 8	2
16 + 10	26
50 + 10 + 10	70
8 - 7	1
10 + 8	18
70, 50, 30, ?	10
20 + 24	44
20 + 30 + 38	88
3 x 3	9
22, 44, ?, 88	66
11 x 5	55
10 x 10	100
80 - 4	76
Half of 40	20

PUZZLE 29

Colour by Numbers

MATHS PROBLEM	ANSWER
RED	1
GREY	2
ORANGE	3
LIGHT BLUE	4
BLACK	5
YELLOW	6
WHITE	7
GREEN	8
PINK	9

Dear Readers,

Thank you for taking the time to read our book.

As a small independent company your support immensely helps us and encourages us to pursue new work. We would be grateful if you can leave a review on Amazon, Goodreads or any other forum that you think would be helpful in spreading the word and helping other readers decide whether or not to read the book. We are always excited to hear from anyone who would like to collaborate on a book. Don't forget to email us any feedback at *learningthroughfun1@gmail.com*.

- Publications in this Range -

Learning Through Fun - Maths Dot to Dot & Other Fun Activities Year 2-3

Learning Through Fun - Maths Cooking Year 2

Learning Through Fun - Maths Cooking Year 3

Learning Through Fun - Maths Cooking Year 4

Learning Through Fun - Maths Cooking Year 5

www.ingramcontent.com/pod-product-compliance
Lightning Source LLC
Chambersburg PA
CBHW051214290426
44109CB00021B/2458